AUTOMATION IMPACTS ON INDUSTRY

A technician at one of the world's largest oil refineries keeps a close eye on energy consumption with a computer. The computer's keyboard console provides the technician with fingertip control of more than 700 refining activities — saving energy and improving productivity (courtesy Honeywell Corp.).

AUTOMATION IMPACTS ON INDUSTRY

By

ROBERT P. OUELLETTE
LYDIA W. THOMAS
EDWARD C. MANGOLD
PAUL N. CHEREMISINOFF

ANN ARBOR SCIENCE
THE BUTTERWORTH GROUP

PREFACE

This study was begun with the ambitious objective of determining the impacts that computers and robotics have had on selected American industries. First, a broad survey of several companies would identify the industries that have shifted to automation and the effects this shift has had on efficiency, productivity, energy consumption, worker performance, reliability, and health and safety. In addition, the study would sort out the effects of automation by focusing on selected plants or industrial sites for more detailed analysis.

It soon became apparent, however, that efforts to meet this objective would be frustrated by a number of obstacles, not the least of which is the pervasiveness of automation in the industries surveyed. It is such an established part of doing business that very few case studies are well documented. In addition, automation is rarely introduced alone in a plant; it is part of a modernization, upgrading or process change exercise. This makes it very difficult to ascribe specific benefits or gains to automation alone.

Nonetheless, this book presents as comprehensive an overview as possible of the use of computer automation, focusing particularly on 10 industries: motor vehicles, aerospace, electronic components, metals, chemicals, glass, pharmaceutical products, foods and beverages, pulp and paper, and water distribution and sewage treatment.

Automation is also an important part of several industries that are not examined here and are outside the "industrial" classification scene. Among the most important are: retail trade; finance and banking; energy exploration, generation, dispatch, transport and distribution; transportation and navigation; and communication.

Automation is generally defined as the replacement of human labor by machinery. In this study the selected area of interest is the replacement of human supervision of machines and mechanized processes by automatic supervision. Automation as addressed in this study is based on modern electronic, digital computer technology to perform monitoring, control

v

and action functions. An older technology, termed "hard automation," is the building of special-purpose machines to perform specific tasks. The topic of hard automation is not covered.

The list of industries potentially using automation was compiled from standard government industrial classifications. A computer-based literature search was performed to collect papers describing examples of automated processes. Additional information was obtained by telephone and mail contact with computer manufacturers, trade associations, government agencies, authors of papers and contacts in industry. Information on current activities was obtained by reviewing the papers presented at conferences and from a visit to an industrial trade show. Personal interviews were conducted with representatives of General Motors and Chrysler Corporations, as well as numerous manufacturers in the machine tool and related computer design fields.

Specifically, examples of the application of automated processes or machines were sought. Descriptions of both the traditional and automated processes were obtained when possible. The most important type of information collected was quantitative measures of the results of automating the process.

The examples presented were selected on the basis of availability of quantitative measurements of effects. Secondary considerations were the desire to include industrial applications that could be used, and the degree of electrification and energy savings that could be achieved by adopting automated computer-based control.

It was difficult to find detailed examples of the results of automation in the open literature. Descriptions of automated processes are numerous, but quantitative measurements of results are either generalizations or not stated in the articles. Frequently, direct contact with the corporation revealed that the results were considered proprietary and would not be revealed, or would only be revealed when an atmosphere of confidence was established.

Information has been obtained on the following effects of automation:

- Energy savings: process industries may be operated at the optimum conditions for energy conservation. Computers can be used to operate manufacturing equipment in the most efficient manner and shut off electrical equipment when not in use.
- Material savings: computer control of equipment such as blast furnaces yields the maximum amount of product from the raw materials consumed. Automated manufacturing operations can reduce scrap losses and maximize efficient use of the feedstock and raw materials.
- Labor effects: automation has frequently been used to replace humans in tasks that are monotonous, physically tiring, dangerous or harmful

to health. Low-skilled workers are replaced by machines, but higher-skilled personnel are required to program and operate the automated equipment. The effect on the labor force must be determined for both the number and skill level required.

- Productivity: an increase in productivity commonly occurs when an existing facility is automated, or when a completely new facility, which includes automation, is compared with a facility using a traditional technology.
- Economics: the savings that result from energy and materials efficiency, reduced labor force, and higher productivity must be balanced against the capital cost of the equipment and the more expensive maintenance.
- Environmental effects: industries that have problems meeting environmental regulations may use computer-controlled processes to minimize pollution emissions.
- Worker safety and health: automation is frequently used for tasks in which the worker would be exposed to hazardous conditions or toxic fumes, as in welding and spray painting operations.
- Product quality: automated plants frequently produce a more uniform product that is consistently of a high standard of quality.

The authors wish to gratefully acknowledge Electricité de France of Paris for the sponsorship and support given for this project. Also acknowledged is the support provided by Ms. Dreamer Hale, Ms. Malinda Miller and Mr. Mario Muradaz.

Robert P. Ouellette
Lydia Waters Thomas
Edward C. Mangold
Paul N. Cheremisinoff

Ouellette   **Thomas**   **Mangold**   **Cheremisinoff**

Robert P. Ouellette is Technical Director of the Environment Division of the MITRE Corporation. Dr. Ouellette has been associated with MITRE in varying capacities since 1969 and has been Associate Technical Director since 1974. Earlier, he was with TRW Systems, Hazelton Labs, Inc., and Massachusetts General Hospital. He graduated from the University of Montreal and received his PhD from the University of Ottawa. A member of the American Statistical Association, Biometrics Society, Atomic Industrial Forum and the NSF Technical Advisory Panel on Hazardous Substances, Dr. Ouellette has published numerous technical papers and books on energy and the environment. He is co-editor/author of the comprehensive Electrotechnology survey series published by Ann Arbor Science.

Lydia Waters Thomas is Associate Technical Director of the Environment Division of the MITRE Corporation. Before joining MITRE in 1978, Dr. Thomas held positions with New York City's Bureau of Hospital Care Services and the Portsmouth, Virginia, Public School System. She obtained her PhD from Howard University and is currently serving as a member of the Environmental Advisory Board to the Chief of Engineers, U.S. Army Corps of Engineers, and Chairman of the Chemicals Regulation Task Group of the U.S. National Committee/World Energy Conference.

Edward C. Mangold holds a PhD in Physics from Oklahoma State University. He spent four years with the U.S. Environmental Protection Agency Enforcement Division in field studies at several power plants and industrial facilities. For the last four years at the MITRE Corporation, Dr. Mangold has worked to improve management control systems used by the Department of Energy for environmental and safety problems in nuclear, petroleum and fossil energy technologies, and has engaged in

ix

automation studies and research. He is co-author of *Coal Liquefaction and Gasification Technologies,* published by Ann Arbor Science.

Paul N. Cheremisinoff is Associate Professor at the New Jersey Institute of Technology, and a chemical engineer with a background in industry and academe. He is a consulting engineer and has been a consultant on environmental/energy/resources projects for the MITRE Corporation. He holds engineering degrees from Pratt Institute and Stevens Institute of Technology, is a member of Sigma Xi and a Fellow of the New York Academy of Sciences. Mr. Cheremisinoff is author of many publications, including several Ann Arbor Science books, such as *Pollution Engineering Practice Handbook, Carbon Adsorption Handbook* and *Industrial Respiratory Protection.*

CONTENTS

FIGURES

TABLES

CHAPTER 1

OVERVIEW OF THE AUTOMATION FIELD

The primary emphasis of this study is the application of process control computers in heavy industry. To place process control in the context of the broad field of automation, this introductory section discusses three subjects:

1. automated machine tools [focusing on numerical control (NC), especially digital numerical control];
2. the growing role of industrial robots; and
3. the use of computer-aided design (CAD) and manufacturing (CAM).

AUTOMATED MACHINE TOOLS

Economic and social considerations encourage adoption of advanced industrial automation. Where advanced technology can increase labor productivity, it can reduce the effect of labor costs on the price of a manufactured product. At the same time, automation can meet social needs by permitting machines to perform unpleasant jobs. This would alleviate growing dissatisfaction among workers who perform jobs that are harmful, potentially dangerous, strenuous or dull [Greene 1981a,b,c; Larsen 1980; Nitzan and Rosen 1976].

These economic and social considerations make automation seem desirable in machining operations where metal is cut and shaped in processes that are repetitive, dull and potentially dangerous. Such operations use machine tools that include lathes, grinders, boring mills and metal-forming presses to shape metal by cutting or pressure. Their products are machined in batches that range in size from several units to several thousand units, and in 1976 they accounted for about 67% of the nation's total manufacturing output or 22% of the gross national product (GNP) [Nitzan and Rosen 1976].

The obvious advantages of automation led to the appearance in the 1950s of NC of machine tools. With NC, a program in the form of a series of holes in a punched paper tape controls the operating sequence of a machine tool such as a lathe. The British have some 100,000 NC tools in operation; there are 25,000 in Japan and some 60,000 in the United States [Marsh 1980a,b,c].

Computer numerical control (CNC) represents an advance over traditional NC. Instructions are fed to a machine via a microprocessor. The fact that the program resides in a solid state memory allows easy modification of the way in which the machine operates. In this way, a manufacturer can make a range of products on one tool. The calculation capability of the microprocessor also gives greater flexibility to the users.

In the 1960s manufacturers started to integrate NC and CNC by linking them to a computer that provides for greater power of computation as well as for control and communication capability between the operator and the machine tools. About 70 digital NC machines in Japan and 12 in the United Kingdom are in operation.

Some machines can change their tools automatically to carry out many machine operations on the same piece. These machining centers bring additional diversity to the manufacturing process.

Flexible manufacturing systems (FMS) are a further step toward integration. Machine centers are linked together such that a piece being worked on can travel from one machining center to another in direct sequence, under the control of one or more control computers. There are some 50 FMS in the world today.

Adaptive control (AC), or the use of information about a machining process to improve the efficiency of the process while it is taking place, holds great promise [Larsen 1981]. It could have a dramatic effect on per-unit manufacturing costs. The application of AC is now confined to the aircraft and aerospace sectors.

Automated machine tools have displayed several benefits [*Manufacturing Engineering* 1980a–j]:

- high repeatability;
- reduction in the intensity of quality assurance measures;
- greater machine utilization;
- control flexibility;
- less scrap;
- high mean time between failures (MTBF);

- simple maintenance, diagnosis and repair;
- easy reconfiguration; and
- less paperwork.

Table I lists more detailed benefits and savings associated with using NC machining tools [Modern Machine Shop 1980].

Although NC machine tools were introduced in 1954, even today they represent only 2.5% of all machine tools sold and less than 3% of all machine tools in existence [King 1979]. Only in the last several years has the cost of programmable controllers declined to a level that will encourage increased use of automated machine tools.

Table I.   Advantages of NC Machines[a]

Advantage	Anticipated Savings from NC Machine
Improved accuracy	5% of direct labor cost
Reduced cutting tool adjustment (by use of tool offsets)	5% of direct labor cost
Reduced cutting tool change time (change only when dull)	20% of tool allowance
Reduced cutting tool cost (throwaway carbides, more standard tools, less specials)	25% of tool cost
Longer tool life due to optimum cutting speeds and feeds	30% of tool cost
Savings in purchasing (less tools, less paper)	5% of tool cost
Improved tool life due to improved machine performance	20% increased tool life
Reduced cutting tool storage (simpler tooling)	50% of tool crib area
Savings in tool maintenance (cutter grinding)	20% of cutter grinding cost
Less toolroom load due to less tooling required	25% less toolroom required
Lower fixture cost (less needed)	75% of durable fixture cost
Less tool engineering time	30% of tool-process engineering cost
Advantage of family-of-parts concept	20% of tool-process engineering cost
Savings from less tool engineering (tool engineering records, tool drawings, process sheets, etc.)	40% of printing cost
Machine maintenance savings due to improved and simpler designs	25% of machine repair (labor)
Less machine repair parts required	25% of machine repair (materials)

Table I,   continued

Advantage	Anticipated Savings from NC Machine
Less inspection due to improved machine (process repeatability)	30% of inspection cost
Inspection more accurate than manual methods	Actual inspection time can be reduced as much as 80%
Reduced setup time	80% of setup cost
Reduced setup scrap	30% of scrap cost
Reduced scrap due to tool change or adjustment	20% of scrap cost
More running time (80–85% vs 40–60%)	10% of total burden
Control of cycle in hands of management (can be fixed)	10% increased production
Savings in setting and maintaining standards	50% of cost of standards
Power consumption more level due to continuous running	5% of power cost
Reduction of inventory	5% of dollar value of inventory
Savings from storage of less productive material	20% of storage area
Less inventory (less material handling)	5% of material handling cost
Floor space savings due to need for fewer machines	Actual space saved
Savings in supervision	Actual number saved
Lower fringe costs due to more productive time	25% reduction in fringe cost
Ability to produce samples with production runs	50% of sample cost
Availability of samples	Useful sales tool
Opportunity for foreman to concentrate on use of people rather than machines	Improved total operation
Reduction of direct labor	Actual savings based on pieces per week — not cycle time
Flexibility of scheduling	Improved customer service
Savings in scheduling	Improved flexibility
Ability to handle engineering changes	Simple program change
Ability to handle variable raw material	Less raw material rejections
Ability to produce more complex parts	Machine capability simplifies tooling
Product engineering has more design flexibility	Can take advantage of NC capability
Ability to handle future designs without extensive tooling	Program changes only will handle many new designs

Table I,   continued

Advantage	Anticipated Savings from NC Machine
Reduced costs and improved estimating accuracy	Estimates can be dry run of tapes
Skills built into tape programs retained through personnel changes	Tool and process engineers improved by 15%

[a]Adapted from Modern Machine Shop [1980].

Numerous other manufacturing operations can be placed under computer control. Spray painting and electric arc or gas welding are two of the most widely used. The control of heat treating operations and other furnace controls is common. Innovative devices have been introduced, such as computer-driven laser cutters of thin-gauge metal, but most applications are with traditional tools and processes [King 1979].

Despite the potential benefits, industry has been slow to adopt computer automation. To understand the reasons for limited penetration in the United States, and the extremely selective use that has been achieved, a description of some of the industrial engineering principles relevant to the selection of computer automation is presented. Additional factors influencing the rate at which automation is adopted in manufacturing include economics, productivity and pressure from foreign competition.

Size of Production Run

It is convenient in discussion of automation technologies to categorize manufacturing operations into three groups based on the number of items produced.

1. Mass production involves large numbers of items such as ball point pens and frequently uses the techniques of hard automation.
2. Batch production involves units of 50 or fewer items and is currently the object of much attention in the application of automation.
3. Custom production involves only one or a very few items that do not have significant differences. This is the most difficult situation in which to justify economically the adoption of automation.

More than 75% of all U.S. manufacturing is accomplished in lots of batch size or smaller, and it traditionally has not been automated [King

1979]. Many of the companies that produce systems for automation are directing their efforts toward designing and marketing automated equipment for operations with smaller production lot sizes. Even process industries such as steel and chemicals frequently change the specifications of the products from the same facility. The application of automation to custom production is difficult because of the time required to reprogram the computer for each item where a variation is required. In the production of trucks and farm machinery, for example, there are so many variations in auxiliary equipment that it is unusual to find two identical units in sequence. The production operation must be altered from one unit to the next.

Manufacturing Engineering

Two methods of sequencing operations and physically positioning facilities in a manufacturing operation are the transfer line and batch production. If the number of units produced is large and the nature of assembly is suitable, the traditional transfer or assembly line, consisting of a linear arrangement of stations that sequentially perform the production operations, is used. The time required to perform each step is a critical parameter in achieving high productivity, since the longest operation determines the speed of the line. Automating noncritical functions does little to improve the speed of the line.

With batch production, a few items are confined to one location where operations in the sequence of production and assembly are performed. After the operations are completed the entire batch is released to the next location. For many factories using this system automation is not economically rewarding because of the small size of the production run.

Manufacturing Functions

The concept of computerized automation is being applied to many functions besides the obvious one of controlling the machine or equipment producing the product. Computers can be used in a wide variety of manufacturing operations. In fact, many workers commonly seen on the manufacturing floor would disappear from a completely automated factory, although some would be replaced by computer operators and programmers.

This section discusses manufacturing functions and steps that can be automated. Very few facilities in the United States use all of the automated steps mentioned below.

Warehousing and Inventory

The first step in the manufacturing process that can be automated is the record-keeping related to the receipt and storage of raw materials and parts. Associated with this is keeping track of goods in process as a part of manufacturing production control. The use of computers with distributed terminals to keep track of stock greatly reduces the human effort required to perform the same task manually and can provide a more up-to-date, complete and accurate inventory.

Material Conveyance

If the inventory is computerized, it is easy to transfer material from storage and between production locations without handling by operators. Conveyer belts, pallet mounting systems and overhead cranes are some of the means of transfer used to move parts under computer control. Considerable savings in labor can result, as well as fewer mistakes and more timely arrival at the destination.

Inspection and Testing

Automated inspection systems range from a simple means for gauging the dimensions of a machined part to elaborate devices that measure the performance of complex products such as automobile engines. In addition to performing tests and recording data, the inspection system often will perform statistical analyses on data collected from the tested parts to determine statistical parameters required to control production tolerances. In addition, the automobile industry is coping with stricter requirements to control exhaust emissions and improve fuel economy that have led to more complicated production line test procedures. A valve used in the pollution control system for a Ford automobile engine is tested with an automated standard that checks 13 functions on every valve.

Data from inspection have been used to correct problems in production equipment. A common use is to measure increasing dimensional tolerances that signal cutting tool wear and the need for replacement. Alteration of reaction conditions in a chemical process based on analysis of analytical data in real time is also commonly found.

Equipment Reliability and Maintenance

Proper maintenance of production equipment can consume much of the time and effort of the production workforce. Breakdowns are costly in terms of lost production time, and production equipment deteriorated beyond performance limits reduces the quality of the product. Automated equipment often can sense and record parameters that are useful in determining proper operation of the equipment and aid in diagnosing maintenance problems.

The measurement of the torque of an electric motor driving a cutting tool is a good indicator of performance. Heat exchanger fouling can be discovered by monitoring temperatures and flowrates. The efficiency of refrigeration equipment in grocery stores has been monitored by computers recording the running time and comparing it with freezer temperature measurements that indicate the load. A loss of efficiency signals a need for maintenance. The amount of electric power or steam consumed per production unit is often a good indicator of system performance. Automatic monitoring of these parameters reveals trends that signal the need for maintenance. To reduce the time required to diagnose malfunctions, automated controllers are frequently equipped with diagnostic software and displays to allow operators to quickly identify and assess problems.

Totally Automated Factory

Automated machinery has been used in both transfer line and batch production operations, but the most effective use of automation may require a totally new method of structuring the manufacturing process. The automated machine must be kept working at its maximum production rate to justify the large capital investment. The speed of the manufacturing operation is frequently determined by the full production speed of the automated machine. Consequently, other functions, such as feeding parts to the machine for processing and quality control inspections, must be adapted to the requirements of the automated machinery. This results in equipment and procedures designed around a small number of very fast automated production machines. This type of organization cuts across traditional lines of organization and authority in a manufacturing operation. An obstacle to more widespread automation is presented by the human relations problems inherent in the required new organizational structure.

INDUSTRIAL PROGRAMMABLE MANIPULATORS (ROBOTS)

One specialized technology currently receiving attention that developed from the automation field is the programmable manipulator. While drawing on the same technology as the NC machine tool, the application of the technology to handling parts and tools has created a new machine that is often directly substituted for human labor. The industrial manipulator is popularly known as a "robot," a name with emotional connotations that interfere with perceiving the proper application of the device in an automated process.

The Robot Institute of America (a trade organization devoted to expanding their use) gives the following definition of a robot:

> A robot is a reprogrammable, multi-functional manipulator designed to move material, parts, tools or special devices through variable programmed motions for the performance of a variety of tasks.

Other factors frequently mentioned in the definition of a robot are five to six degrees of freedom in positioning and a gripping device, or "hand" on the end of the arm suitable for grasping the intended part.

Robots are commercially available in a large variety of sizes, weight-handling capacities up to 1000 kg, simple to very complex and with a variety of control systems. A computer- or microprocessor-based control system with an electric or hydraulic servomechanism for activation is the type most frequently used in automation.

Although robots have been available for 20 years, their numbers increased slowly in the early years. Only in the last five years, with the availability of computer technology control systems, has their use greatly increased. The number of robots in use in the world is difficult to estimate. Table II is the best available estimate as of 1980. Japan is the dominant country, followed by West Germany and the United States. Of the up to 7000 robots currently employed in American industry, more than 55% are employed in building automobiles [Engelberger 1980]. Figure 1 indicates that this share of production is expected to continue to 1990 as robot production increases. Unimation, Inc., the largest American robot builder, produces about 50 machines per month. Independent studies have predicted an increase of about 35% per year for the current decade. This would result in a total number of robots in the American automotive industry of about 50,000 in the year 1990.

Experts have identified 20 technical attributes of robots [Heer 1980]:

1. work space command with six infinitely controllable articulations between the robot base and a hand extremity;
2. fast hands-on, instinctive programming;
3. local and library memory of any size desired;
4. random program selection by external stimuli;
5. positioning repeatability to 0.3 mm;

Table II.   Estimates of Worldwide Distribution of Robots [Engelberger 1980]

| | Programmable Robot Type | | | | | |
| | Servo-controlled | | Nonservo-controlled | | | |
Nation	Point-to-Point	Continuous Path	General Purpose	Die Casting And Molding Machines	Mechanical Transfer Devices (pick and place)	Total
Japan	3,000[a]		11,000[a]		33,000	47,000
West Germany	1,800	345	1,100	[b]	5,000	5,850
United States	60	85	20	20	[b]	3,255
Britain	70	40	200	50	[b]	185
Poland	7	3	[b]	3	360	720
Belgium	450	120	[b]	[b]	7	20
Sweden	20	20	100	30	[b]	570
Norway	20	10	70	10	30	200
Finland	300	150	200	200	20	130

[a] Total number of both types.
[b] Figure unavailable.

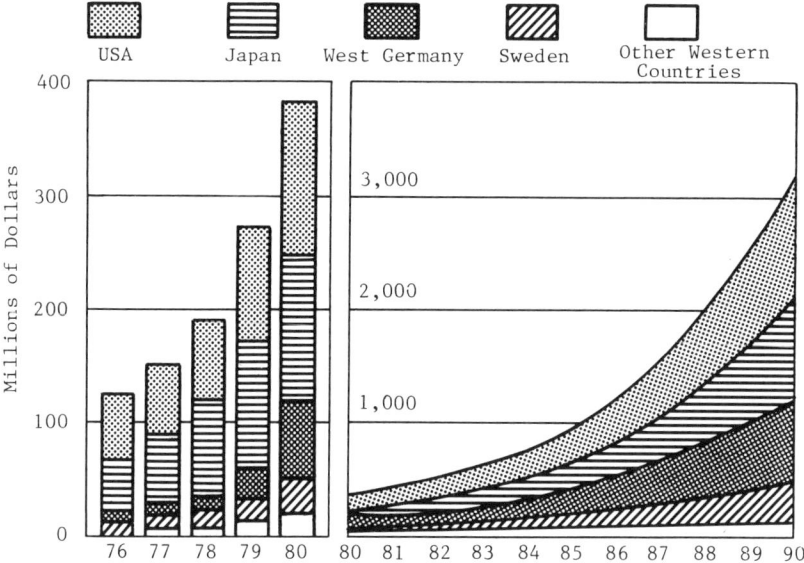

Figure 1.   Robot market trends and projections [Heer 1981].

6. weight handling to 150 kg;
7. point-to-point and path-following control;
8. synchronization with moving targets;
9. compatible computer interface;
10. high reliability (at least 400-hr MTBF);
11. rudimentary vision orientation and recognition;
12. tactile sensing for orientation, physical interaction and recognition;
13. multiple-appendage hand-to-hand coordination;
14. computer-directed appendage trajectories;
15. mobility;
16. minimized spatial intrusion;
17. energy-conserving musculature;
18. general-purpose hands;
19. voice communication; and
20. safety.

Robots are commercially available with the first 10 characteristics. The next set of desirable attributes should be integrated in robots before 1984 [Engelberger 1980].

Robots are typically used in the United States for [Wallace undated]:

• spot welding,
• arc welding,

- forging,
- plastics,
- machine tool loading and unloading,
- material handling,
- investment casting, and
- assembly.

Robots could easily make [March 1980a,b,c]:

- pumps,
- compressors,
- air blowers,
- speed-change gears,
- calculating machines,
- typewriters,
- knitting and sewing machines,
- fire extinguishers,
- freezers,
- gas and oil appliances,
- ventilators,
- vacuum cleaners,
- refrigerators,
- washing machines,
- lighting fixtures,
- radio and television sets,
- semiconductors, and
- car parts.

Important research and development work is underway to develop robots with the following capabilities [Wallace undated]:

- small parts assembly,
- line-tracking,
- painting,
- computer hookup or direction, and
- visual feedback.

The popularity of robots in industry will be affected by their declining cost, technological improvements, and the impending development of new characteristics and capabilities such as vision [*Optical Spectra* 1981], as well as the advent of microprocessors [McCool 1979].

The reasons why industrial robots have not been accepted at a faster pace have been summarized by Clapp [1979], who studied several case histories of failures to successfully use robots. The author concluded that the failures are associated with human factors and organizational problems, not with technical problems. They are due to the reactions of both management and labor to the introduction of new technology into the very conservative industrial environment. Clapp and the authorities he quotes consider the opportunities for technically and economically successful implementation to be numerous but minimally implemented, blocked mostly by management caution. Installation is favored for dull, monotonous, dangerous or unpleasant tasks, where the workers appreciate being relieved of the work and accept the robot. However, there is concern when the robot disrupts traditional patterns of organization and management. An example was given of the industrial plant where the robot so disturbed the organizational structure that it was replaced. Clapp concluded that the acceptance of robots in factories will continue at a gradual pace.

Robots are bought by industrial manufacturers to reduce cost, improve productivity or for noneconomic justifications, including [Ernst 1980]:

- health and safety;
- higher quality of workmanship;
- minimum downtime—1-2% for a robot compared with 9% for a factory worker;
- reduction of scrap and rework;
- consistency and reliability; and
- flexibility—robots can easily be moved to another assignment.

Economics remains the basis of most decisions, with manufacturers often discouraged by the large capital investment required for robots. An average payback time of no less than seven years is a formidable limitation to their expanded use [Koekebakker 1980]. Nonetheless, robots cost less than wage-earning workers over the long term. The hourly cost of an industrial robot in 1979 stood at about $4.50/hr and increases are projected to be very small over the next few years. On the other hand, the hourly rate of a typical worker is growing yearly in a semiexponential fasion (Figure 2) [Engelberger 1980].

Analyses of the economics of robotic assembly lines take the form of Figure 3, showing the relationship between cost and production volume for assembly done by robots, fixed automation and humans [Sugarman 1980]. The actual curves will be different for each case, but the general relationship will prevail, showing the economic advantages of fixed and programmable automation over human labor in the volume range of 0.3-10 million units [Allan 1979].

Robots have been identified as possibly promoting a 25% increase of production. The forecast assumes that robots will take over 5% of the blue collar jobs in the United States in the next 50 years [Munson 1978], an impact much smaller than previously anticipated.

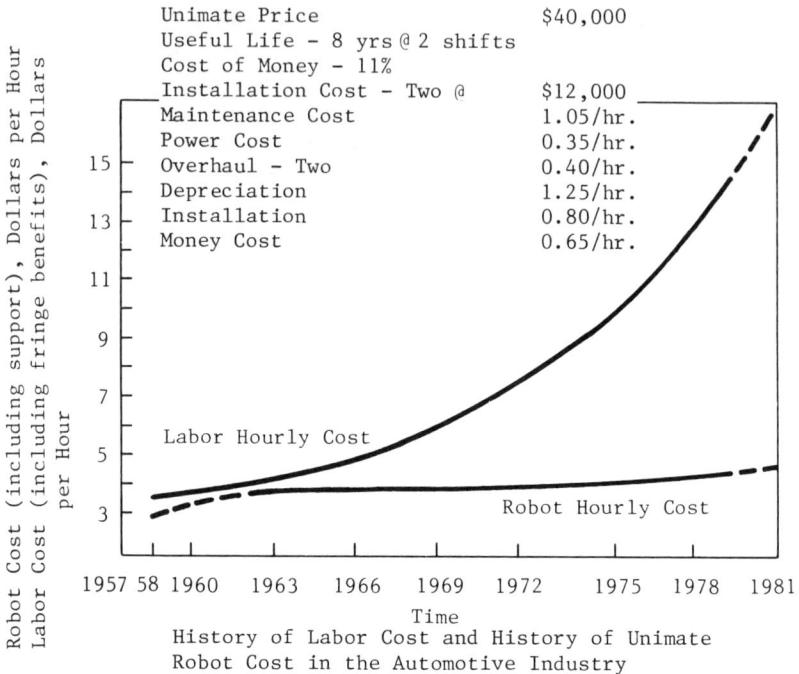

Figure 2.   Robot cost assumptions as of 1979 [Engelberger 1980].

COMPUTER-AIDED DESIGN/MANUFACTURING

Computer graphics can be used in designing, drafting and analyzing through the dual systems of CAD and CAM. A wide acceptance of CAD/CAM can be at the heart of a revitalization of the manufacturing industry by spurring increases in productivity [Bylinski 1981]. The systems are already having a favorable impact on prices, productivity and product quality in the United States. Optimistic engineers claim that (1) labor can be reduced by 5:1 to 6:1 with CAD; (2) lead time can increase by 2:1; and (3) these ratios go as high as 30:1 and 50:1 when CAD is linked to CAM [Bylinski 1981].

CAD/CAM is used by makers of electrical components, aircraft and farm machinery. However, few manufacturers use the full-fledged CAD/CAM capability. For the most part, only disconnected blocks of the CAD/CAM complex are used. CAD/CAM can be awesome in its complexity and often forbidding to all but the larger manufacturing companies.

The CAD/CAM business is growing. Sales totaled $750 million in

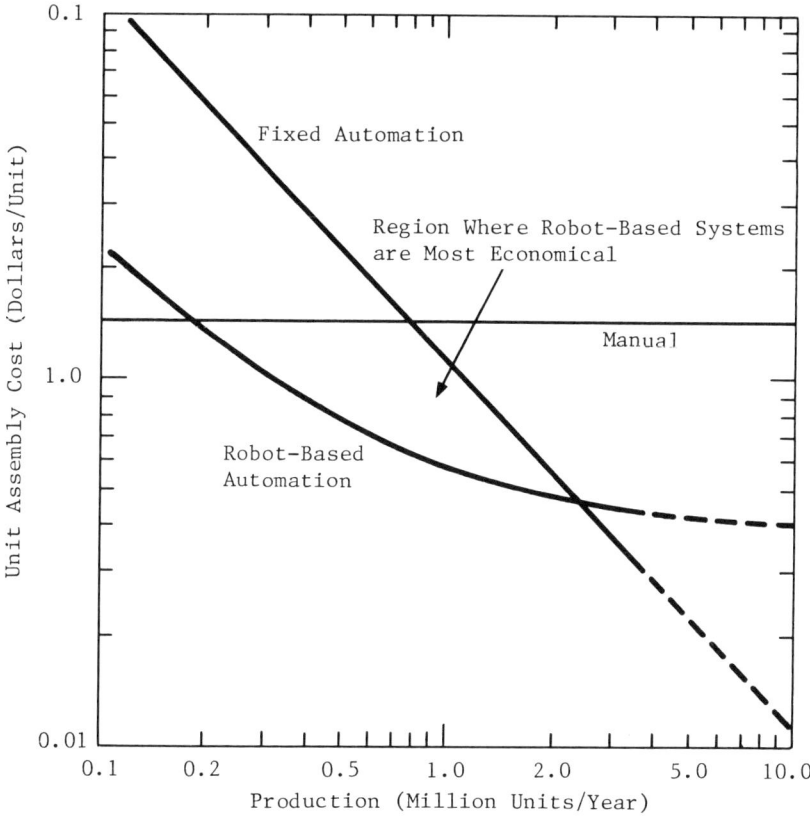

Figure 3.   Economics of automation [Sugarman 1980].

1981 and they are expected to reach $4.4 billion in 1984, with an average growth of 41%/yr. However only 3500 CAD/CAM systems have been installed out of a possible 30,000 sites.

Computer-Aided Design

The traditional approach to the design of parts to be manufactured is to produce a mechanical drawing showing three dimensions—a highly labor-intensive process. These drawings, together with process information and standards cited by reference, are the formal documentation governing production and inspection of the part. Often a computer can assist both the design engineer in performing calculations of mechanical stress or thermal properties required to design the part and the draftsmen in preparing the drawings.

The CAD method replaces the traditional drafting board with a special computer console on which the part is illustrated on a cathode-ray tube (CRT). A rapid method of altering the design, such as a light pen, is used to change the dimensions of the part being designed. The calculations used in producing the part are included in the computer program along with instructions necessary to perform the graphics operations. The final result is a part that is defined and illustrated in numerical form by a computer. Subsequent processes, such as production planning, can also be completed using computerized systems rather than by traditional methods.

One principal advantage of CAD systems is that they are much faster than traditional drawing board methods. This increased speed is noticed in the original design and in the numerous revisions that must be made to many components as the product evolves.

Computer-Aided Manufacturing

The use of computer technology in manufacturing operations began in 1954 when NC machine tools were introduced. Because of the large amount of time required to prepare the punched paper tape used for control, NC machine tools were only used in a few large shops. In the past few years, their use has expanded as better and less expensive equipment has become available to prepare machine instructions. As noted above, the modern trend is to direct numerical control (DNC), in which the computer is linked directly (electronically) to the machine tool. This is superior to the older systems in which the computer produced a punched paper tape that was read by the machine tool. Most modern machine tools use a microprocessor with a solid state memory as the control element; the microprocessor itself can be linked to a higher-level computer for DNC, or it can operate independently.

Integration of Computer-Aided Design and Manufacturing

The largest benefits to be obtained from automation will be achieved when the design and production operations are integrated into a complete manufacturing system. Some examples of this level of integration are:

1. use of CAD of the component to produce the numerical control program that operates the machine tool;

2. use of the original CAD specifications as the reference for quality control inspection;
3. use of part design dimensions to fabricate a stamping die;
4. use of the design information to perform production planning;
5. automatic detection of defects in the programs for parts production before use is attempted; and
6. use of a uniform set of dimensional references to ensure a better interface between parts.

Many manufacturing operations can be computerized to form a totally automated plant. In addition, all of the production control and accounting operations that accompany manufacturing can be included in an automated system. The result will be a totally automated factory that differs in many ways from today's factory. A totally automated factory will have a smaller workforce and a smaller number of automated machines. The physical layout of the facility and the flow of items through the production cycle will differ noticeably from those in present practice. Some of the components of a completely automated factory are being developed today, but another decade will be required before all of the components are perfected and the concept is widely practiced. Flexible manufacturing is making progress in U.S. industry [Hutchinson 1979; Lerner 1981; Stauffer 1981]. The most advanced case can be seen at the Yamazaki machinery work in Noagoya, Japan [*Time* 1981].

National Standards Programs

The CAD/CAM field is just beginning; consequently, no generally accepted standards exist to define how the tasks arc to be performed, and each developer is free to establish his own. Because of the lack of standardization, many different systems exist, each developed to meet the needs of an individual user. This lack of standardization has resulted in several national programs centered around the aerospace and machine tool industries. They are setting up data processing architectures and methodologies that will ultimately lead to a set of standards to be adopted by all industries.

Potential Impact

The Advanced Technical Planning Committee of Computer Aided Manufacturing International outlined a long-range strategy for the development of an information service to improve productivity in design and manufacturing in the world durable goods industry. They estimate a potential savings of $36 billion (in 1975 U.S. dollars) if CAD/CAM is fully implemented by the durable goods industry of the developed market economies.

PROCESS CONTROL COMPUTERS

History

The literature of digital computers is truly immense and will not be reviewed in detail here. Instead, it will suffice to touch on a few historical highlights.

The first use of digital computers for closed-loop control can be found in aircraft flight control in 1955; first monitoring in electric utility in 1958; first supervisory control in a refinery in 1959; and first use in a chemical plant in 1960 [Kompass 1979].

The cumulative total of digital computers installed in industrial control applications since the 1960s comes to more than 500,000 units. The annual growth rate has been in excess of 65% [Kompass 1979]. Figure 4 shows that some 250,000 new ones will be added this year. Control applications will account for 20% of all computers — but 40% of the mini- and more of the microcomputers.

Taxonomy of Computers

The computers discussed below are either analog or digital. Digital computers are divided into supervisory control and direct digital control. These different categories are characterized in Table III, where their major advantages and disadvantages are presented.

Trends in Computer Control, Sensors and Actuators

The first significant trend in computer technology of the last 25 years has been the reduction in the size of the computer. Machines have shrunk from main frame to minicomputer to microcomputer. At present the microprocessor, a complete logical processing and memory unit on a single circuit chip, is finding widespread applications in the automation field. There has been a correspondingly dramatic decrease in the price paid for a computer to perform a given set of mathematical operations.

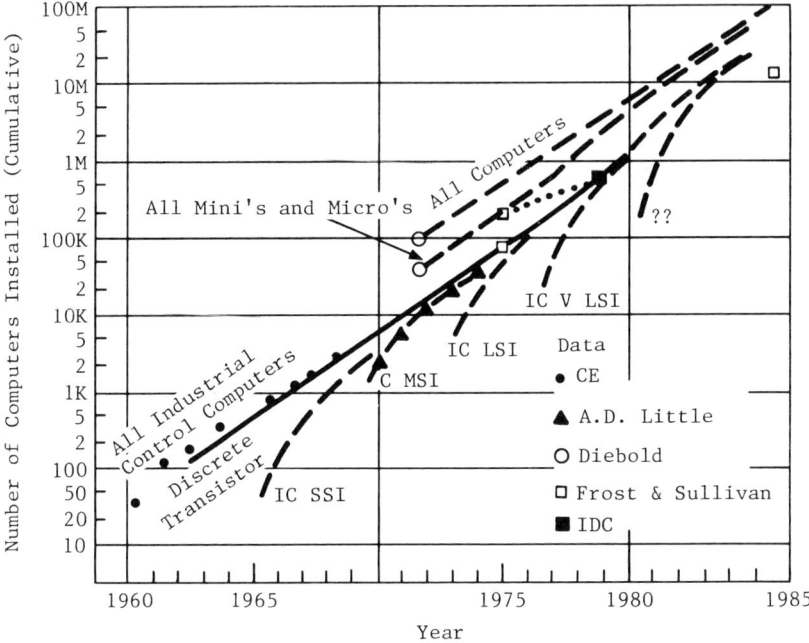

Figure 4.   Projected growth of computers [Kompass 1979].

As computer technology progressed from vacuum tubes to semiconductors to integrated circuits to very large-scale integrated circuits, the cost per million calculations dropped three to four orders of magnitude from 1957 to 1981.

Other characteristics of the computer also changed to make their application to the field of automation more attractive. Computer reliability increased, so that instead of expecting one or more computer breakdowns a day, the plant operator can now reasonably expect a year of operation without a computer system component failure. As memory size increased, the computer's versatility was expanded. Instrumentation and supporting equipment have transformed from analog to digital technology for integration with the computer. All of these trends reduced the cost of the computer technology required for a given automation application.

Other developments increased the ease with which the operators could interact with the computer. Systems began operating in real time rather than in batch mode. Compiler languages were simpler to master and interactive programs from consoles giving rapid input and output were

Table III.   Comparison of Process Control Computers

Category	Characteristics	Advantages	Disadvantages
Analog Control	Analog signals from sensors are carried by wire and pass through signal conditioning steps; system responds with alarm when signals move outside of preset range	Reliable; inexpensive	Not programmable; signal conditioning required; unable to recover from strong transient associated with upsets, shutdown and startup.
Supervisory Control	Calculate and transmit set points or targets to dedicated controllers	Very reliable; moderately flexible; computational capability; can be implemented incrementally	More expensive; slower than DDC
Direct Digital Control Centralized	Performs control function as well as monitoring process conditions, calculating new set points, and correcting errors	Faster than supervisory control; high degree of flexibility	Power outage inactivate entire system; complexity of programming
Direct Digital Control Distributed		More reliable than centralized system; high degree of flexibility; minimize wiring cost; control loops distributed among several computers	

adopted. The availability of technical personnel trained to use computers has been and will remain one of the most significant factors affecting automation technology.

Programmable controllers (PC) are replacing the traditional relay box. They are solid state devices amenable to control sequence programming [Lutz-Nagey 1976]. Meanwhile, display systems are moving from the traditional annunciator lights to color CRT displays supported by sophisticated graphic software.

The trend to minicomputers during the 1960s and microprocessors during the 1970s increased the expansion of process control by computer. The first applications used large computers to control complete plants. As smaller computer systems became available and computer system costs decreased, they could be easily applied to smaller processes. The availability of minicomputers and microprocessors in the $10–100 price range also made distributed computer systems possible, in contrast to concentrating all of the control into one central system. Reliability was increased and implementation could take place at a simpler level.

The future trends predicted are:

- more distributed control systems using microprocessors;
- more use of graphics displays on CRT in control rooms;
- separate computers to independently perform signal processing, and computers and display in parallel;
- digital processing in distributed units replacing analog processing;
- control software that is easier to use; and
- use of total plant simulation and process analysis during plant design.

Sensors

The two most important types of measurement are temperature and pressure. Flowmeters of the electronic type have appeared on the scene. Timing, counting, force, torque, acceleration, position, pH, moisture and density are also often measured in process control applications. The capability of transducers is nowhere more evident than in laboratory and analytical instruments.

Actuators

Process control valves, electric motors and drives, hydraulic servos and fluid power are the most frequently used actuation devices.

Rationale for Using Computers and Results

Computers are used to perform four functions: process monitoring, analysis, control and optimization. It is interesting to review management objectives in opting for automation (Table IV). Raw material and fuel savings are paramount among the objectives. The use of computers for process control reduces energy consumption in four ways:

1.   Better unit management saves energy in the process.
2.   Energy is saved when a greater amount of product is produced from a given quantity of raw material.
3.   Energy savings result from better quality control and increased production from a facility.
4.   Indirect energy savings result from improved process and equipment design made possible by computer control.

Figure 5 illustrates the results of a survey conducted between 1961 and 1977 that included 34 plants in the chemical, petroleum, paper, glass,

Table IV.   Management's Objectives in Introducing Computer Process Control [DOE 1979]

Objective	Number of Times Mentioned At Survey Plant[a]
Reduce Raw Material and Fuel Costs	9
Optimize Production	6
Increase Production	5
Increase Process Knowledge	4
Improve Plant or Unit Operating Efficiency	4
Improve Product Quality	4
Gain Experience with Computer Process Control	4
Reduce Manpower Requirements	4
Provide Better or More Rapid Analysis	3
Increase Equipment Availability	2
Other[b]	5

[a]Several objectives generally were given at each survey plant.

[b]These objectives include the desire to improve data-gathering procedures, reduce equipment damage, improve plant safety, achieve process stability and increase information available to operators.

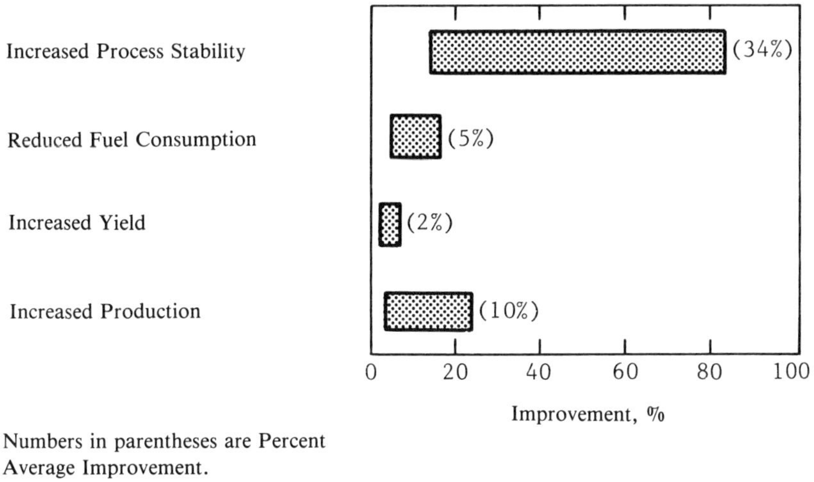

Increased Process Stability   (34%)

Reduced Fuel Consumption   (5%)

Increased Yield   (2%)

Increased Production   (10%)

0   20   40   60   80   100

Improvement, %

Numbers in parentheses are Percent
Average Improvement.

Figure 5.   Industrial improvement through computer application [DOE 1979].

cement, iron and steel industries. Automation brought dramatic improvements in process stability, energy conservation, yield and production, as
shown in Figure 5. Note that these results apply to the situation before
the energy crisis of 1974 [DOE 1979].

CHAPTER 2

EXAMPLES OF INDUSTRIAL EXPERIENCE WITH AUTOMATION

SELECTION OF INDUSTRIES FOR ANALYSIS

Industrial sectors were selected for review using the criteria:

1. importance of the industry as exemplified by the value of shipment;
2. energy use in the manufacturing process;
3. interest and achievement in automation; and
4. interest expressed by Electricité de France in certain industries.

Table V provides integrated statistics for 1977 (the latest year for which census statistics are readily available). Energy use by the major industry groups is given in this table. Energy derived directly from in-plant combustion of fuel is separated from energy supplied to the production process via generation of electricity. These industry groups are ranked in the table according to their total energy use. This includes a small portion of electricity generated at the manufacturing site rather than supplied by a public utility.

The following industries were selected for more detailed analysis:

1. motor vehicles,
2. aerospace,
3. electronic components,
4. metals,
5. chemicals,
6. glass,
7. pharmaceutical products,
8. foods and beverages, and
9. pulp and paper.

Table V. Purchased Fuels and Electricity Consumed by Major Industry Groups in the United States (1977) [DOC 1980a–f]

SIC[a]	Major Group	Rank	All Fuels Btu×10¹²	All Fuels $×10⁶	Coal and Coke Btu×10¹²	Coal and Coke $×10⁶	Fuel Oil Btu×10¹²	Fuel Oil $×10⁶	Natural Gas Btu×10¹²	Natural Gas $×10⁶	Other, Not Specified Btu×10¹²	Other, Not Specified $×10⁶
20	Food and kindred products	6	952.4	2,537.8	90.1	100.0	195.7	470.5	426.5	743.3	103.5	172.3
21	Tobacco products	20	20.8	63.9	W[a]	W	7.1	16.7	2.3	4.2	7.1	.2
22	Textile mill products	10	339.2	1,136.6	31.6	42.9	102.2	245.4	74.0	134.6	37.2	70.0
23	Apparel and other textile	16	65.6	285.8	1.8	2.8	8.7	22.4	15.2	27.0	17.3	4.4
24	Lumber and wood products	13	227.7	726.5	W	W	48.8	130.6	64.0	112.2	59.7	84.4
25	Furniture and fixtures	17	52.7	198.3	2.9	4.3	8.1	21.8	18.0	32.2	9.5	17.9
26	Paper and allied products	3	1,308.4	2,960.7	226.9	260.3	506.2	1,057.6	350.0	590.1	73.2	128.6
27	Printing and publishing	14	92.3	421.7	W	W	8.7	23.6	28.4	53.4	19.2	13.7
28	Chemicals and allied products	1	2,987.5	6,448.7	342.1	380.9	405.1	915.8	1,484.0	2,049.4	247.5	301.1
29	Petroleum and coal products	4	1,303.4	2,445.4	4.8	5.2	83.4	192.8	1,051.0	1,546.3	60.9	27.0
30	Rubber and miscellaneous plastic	11	272.3	962.4	23.9	24.3	65.7	154.7	71.3	124.4	34.4	58.8
31	Leather and leather products	19	23.1	82.6	1.2	1.0	7.4	17.8	5.5	10.0	4.1	8.0
32	Stone, clay and glass products	5	1,251.8	2,587.4	354.1	383.2	163.1	393.5	523.5	890.3	105.6	158.5
33	Primary metal products	2	2,539.4	7,043.6	631.6	1,823.6	332.1	754.5	905.6	1,496.8	131.9	221.8
34	Fabricated metal products	7	395.3	1,348.7	12.4	22.0	49.5	127.9	184.3	348.7	59.0	104.8
35	Machinery, except electrical	9	339.6	1,268.0	21.3	36.9	41.6	107.6	127.9	234.7	51.7	92.5
36	Electric and electronic equip.	12	249.3	986.7	14.6	18.5	31.3	79.4	84.4	158.3	33.8	61.1
37	Transportation equipment	8	389.9	1,402.6	53.7	90.4	51.8	125.4	137.6	262.0	41.1	75.7
38	Instruments and related products	15	78.4	272.5	W	W	12.7	30.7	17.7	32.9	29.1	6.1
39	Miscellaneous manufacturing	18	49.1	199.9	W	W	9.7	24.9	14.7	28.6	10.6	5.8
	All industries		12,929.0	33,379.6	1,844.8	3,241.6	2,138.9	4,913.4	5,585.7	8,879.4		1,830.4

SIC[a]	Major Group	Rank	Electricity Btu×10^{12}	Electricity \$×10^6	Percent Electricity Btu	Percent Electricity \$	Value of Shipment (10^9 dollars)	Energy Intensity (10^3 Btu/\$)	Ratio of Electricity to Total Energy ×10^{-2}	Ratio of Cost of Energy to Value of Shipment	Percent Electricity Generation Onsite
20	Food and kindred products	4	136.6	1,051.6	14.3	41.4	177,357.3	5.67	14.34	14.31	5.48
21	Tobacco products	20	4.3	32.8	20.7	51.3	8,524.8	2.44	20.67	7.50	
22	Textile mill products	9	94.2	643.7	27.8	56.6	39,439.9	8.60	27.77	28.82	1.56
23	Apparel and other textile	15	22.6	200.6	34.4	70.2	39,189.1	1.67	34.45	7.29	
24	Lumber and wood products	13	55.1	365.5	24.2	50.3	38,110.0	5.97	24.20	19.06	1.82
25	Furniture and fixtures	17	14.2	122.2	26.9	61.6	16,162.9	3.26	26.94	12.27	
26	Paper and allied products	3	152.1	923.8	11.6	31.2	50,356.5	25.98	7.13	58.79	38.06
27	Printing and publishing	14	36.0	307.9	39.0	73.0	48,501.0	1.92	39.00	8.75	0.06
28	Chemicals and allied products	2	508.8	2,721.0	17.1	42.2	109,331.8	27.33	17.03	58.98	7.99
29	Petroleum and coal products	7	102.9	639.7	7.9	26.2	93,945.7	13.87	7.89	26.03	14.03
30	Rubber and miscellaneous plastic	12	77.0	600.3	28.3	62.4	38,237.0	7.12	28.88	25.17	1.84
31	Leather and leather products	19	4.9	45.8	21.2	55.4	7,122.3	3.24	21.21	11.60	1.03
32	Stone, clay and glass products	6	105.5	762.0	8.4	29.5	33,780.9	37.06	8.43	76.59	1.44
33	Primary metal products	1	538.2	2,746.9	21.2	39.0	98,900.8	25.68	21.19	71.22	7.94
34	Fabricated metal products	10	90.1	745.3	22.8	55.3	85,779.1	4.61	22.79	15.72	
35	Machinery, except electrical	8	97.1	796.3	28.6	62.8	115,510.3	2.94	28.59	10.98	6.97
36	Electric and electronic equip.	11	85.2	669.4	34.2	67.8	83,002.0	3.00	34.17	11.89	
37	Transportation equipment	5	105.7	849.1	27.1	60.5	161,510.0	2.41	27.10	8.68	
38	Instruments and related products	16	18.9	164.3	24.1	60.3	26,949.1	2.91	24.11	10.11	
39	Miscellaneous manufacturing	18	14.1	126.7	28.7	63.4	18,077.7	2.72	28.72	11.06	
	All industries		2,263.5	14,514.9	17.5	43.5			17.51		

[a] Standard Industrial Classification Code.

W = Withheld

HISTORICAL TRENDS

In the selected industries, as in virtually all industrial sectors, the same evolution can be found. Digital computers were originally introduced to support business-oriented operations. Accounting, inventories, payroll, maintenance and production schedules, orders and invoices, and sales forecasting were soon automated in many factories.

Engineers and scientists discovered the computational capability of digital computers and used them increasingly to solve complex mathematical problems involving simultaneous equations, differential equations, linear programming and simulation.

The next logical step in the use of computers was as a process controller used in real time and on-line with instruments monitoring status and process variables. Process computers evolved starting with simple data loggers collecting operational data and setting up alarms when critical conditions were met. On a typewriter-like device they printed summary information throughout the day.

A more sophisticated approach followed — sequence controlling. This is an open-loop system. The operator feeds information to the computer on observing the process. Then the computer performs mathematical and logical analyses to adjust process conditions. It directly manipulates process variables and parameters to meet desired conditions and achieve a desired result based on previously programmed steps.

The logical extension of sequence controlling was the supervisory process control system. This is a closed-loop system in which an operator is not required under normal conditions. The computers: (1) receive information directly from instruments monitoring the process (analog signals are converted to digital signals through an analog/digital converter); (2) set control points (fed in the computer by the operator under the previous concept); (3) perform computations; and (4) send signals to the operation to adjust control variables.

Supervisory control can direct the analog controller by digital sequences. In many computers, a full direct digital control is used. Special-purpose computers, especially mini- and microcomputers, are increasingly dedicated to process control tasks or even incorporated into the process or the monitoring instrument.

MOTOR VEHICLES

Industry Characteristics

This section discusses three segments of the motor vehicles and equipment industry as categorized by Standard Industrial Classification (SIC) codes of the U.S. Department of Commerce (DOC): motor vehicles and car bodies (SIC 3711), truck and bus bodies (SIC 3713), and motor vehicle parts and accessories (SIC 3714).

The motor vehicles and car bodies industry comprises establishments primarily engaged in the manufacture of passenger cars, passenger car bodies, trucks, buses and wheeled combat vehicles. Complete car chassis and complete truck chassis, with or without cabs, are also included in this industry. The total value of shipments for establishments classified in the motor vehicles and car bodies industry amounted to $76.2 billion in 1977 — an increase of 78% over 1972. In 1977 the industry had 343,600 employees — an increase of 1% over 1972.

The truck and bus bodies industry is comprised of establishments primarily engaged in the manufacture of truck and bus bodies, sold separately or for assembly on purchased chassis. The total value of shipments for establishments classified in this industry in 1972 amounted to $2.00 billion, and the industry had 34,800 employees.

The motor vehicle parts and accessories industry comprises establishments primarily engaged in the manufacture of parts and accessories for cars, trucks and buses. The total value of shipments for establishments classified in this industry amounted to $35.75 billion in 1977 — an increase of 95% over 1972. In 1977 the industry had 450,700 employees — an increase of 12% over 1972.

The automobile industry is considered the outstanding example of mass production technology. The number of items produced place it in the mass production category, but some of the production runs of components are of batch lot size. The production technology is complex and the industry has a large income in comparison to other U.S. industries. The auto industry has been a leader in the use of hard automation for decades. Both management and labor recognize the need for high productivity and adopt new techniques to achieve it.

The auto industry is experiencing dramatic changes: production is declining, profits are vanishing, prices are going up, cars are becoming smaller and customers are keeping them longer. Because the industry is experiencing major changes in vehicle design to achieve fuel economy, rapid changes in manufacturing methods are also occurring. With a large investment in machine tools currently being made, today's trends in the auto industry will heavily influence the types of automation used in other industries for the next few years.

Computer-aided design (CAD) has been developed by the auto industry to perform the numerous iterative design changes required to optimize an auto body design and minimize weight. The design is now beginning to be integrated with the manufacturing process. The automobile industry uses more than half of the industrial robots used in the United States.

The next significant development will be utilization of robots to perform light pick and place assembly operations.

Computer-Aided Design and Manufacturing

The use of CAD and computer-aided manufacturing (CAM) in the American automobile industry is rapidly increasing. Automobile manufacturing is highly labor-intensive, with American labor receiving the highest wages paid in the manufacturing industry. The large production volume (6 to 8 million vehicles per year) justifies the use of automated production methods. As noted above, the automobile industry has long been a leader in the use of hard automation—special tools built to perform one function repetitively. Today the automotive industry is also beginning to take the lead in adopting computerized production technologies, using robots for welding and spray painting, CAD for automobile bodies and numerical methods for other manufacturing functions.

These changes in manufacturing are being accelerated by the rapid changes in automobile design caused by the need for higher gasoline mileage and reduced air pollution emissions. A complete new generation of cars is being introduced by the "Big Three" U.S. automobile manufacturers. Completely new components, engines, transmissions and bodies, which are not derived from any previous designs, must be produced. The most convenient time to introduce new technology into the manufacturing process is when a totally new production facility is established. Since so many auto models are being changed, the new equipment being purchased by the auto manufacturers is a dominant force in the market for automated machinery.

The initial response to the requirement for weight reduction was to retain the interior car dimensions and shorten the exterior dimensions. A savings of 700–800 lb (318–364 kg) resulted. The next step was to redesign new body shapes with front-wheel-drive systems. The present cars will require new assembly lines for all components. By 1985 additional weight savings will be achieved as aluminum, plastics and composite materials replace steel, and as new production and bonding technologies are employed. The automobile industry invested $30 billion in production technology to reduce weight by 800 lb (364 kg) between 1973 and 1978. Another $30 billion will be required to finance production facilities for the next round of weight reduction improvements by 1985, and a large sum will be spent on computer-aided production technology.

The use of CAD in the automobile body evolution was pioneered by General Motors (GM) in a $50 million effort beginning in the late 1950s [Anker-Johnson 1980]. Some results of this program were:

- development of a graphics console;
- interactive computer software;
- mathematical modeling of auto body shapes;
- numerically controlled machining of these shapes; and
- dynamic analysis of auto body structures to the applied stresses.

Automobile bodies are designed on computers with dozens of iterations required before a design is completed and ready to be released to production. General Motors does not send paper prints to a manufacturing division when it is time for an automobile body to go into production. Dimensions of all parts are sent in digital form. Sheet metal applications appeared first, but an even greater potential is expected for components and power trains. GM has 600 design graphics computer consoles now in operation, and the number will expand rapidly during the next decade. Half of the $10 million annual acquisitions are commercially purchased turnkey systems, while the other half require custom design for the GM application.

The greatest benefits from CAD are realized when the design is integrated with the steps that follow in manufacture. The present practice of producing stamping dies using numerical control will expand during the 1980s to include process planning, factory management and programming of robots. Since automotive manufacturing volumes are greater than aerospace production runs, GM expects that the use of CAD/CAM will parallel the aerospace experience, but for production of tools rather than production of product parts.

New plants will feature faster production due to the use of automated equipment [*Business Week* 1980a,b]. At two new GM plants, 150 cars per hour will be produced by the same number of employees that used to produce 115 cars per hour at the 60-year-old plants that are being replaced. A Ford engine plant produces 250 four-cylinder engines per hour, compared to 200 per hour at the best eight-cylinder plants. The new engine and transmission plants are so capital-intensive that they will be used for three-shift production. Significant energy savings and productivity increases commonly occur when a company builds a new automated plant rather than rebuilding an old one. Ford, GM and Chrysler are planning to close 7 out of 46 existing assembly plants.

A difficulty in implementing CAD/CAM in the automobile industry is the lack of a set of uniform standards applying to procedures for parts design and description as well as to the transfer of information among different computer systems used by different divisions and companies. GM divisions and suppliers are consistent among themselves, but communications across organizational lines are difficult or impossible without a uniform national standard for computer-aided graphics transfer.

Robots in the Automobile Industry

More than 55% of the 6000–7000 robots now in U.S. industry are building automobiles, typically with less than 2% downtime [Munson 1978]. This share of the robot market is expected to persist. During 1981 some 700 more robots will be purchased. Assuming a yearly compounded growth rate of 35%, the 1990 production should be around 50,000 robots. Characteristics of the auto industry that favor the use of robots include [Engelberger 1980]:

1. Medium-scale production runs are not large enough to justify dedicated automation.
2. A broad mix of models in the product requires changes in the manufacturing equipment.
3. Rapid product obsolescence may require unplanned changes in production.
4. Human labor is used for tasks that are tiring, disagreeable and repetitive.
5. Multiple-shift operations produce a better return on investment for purchase of a robot.
6. High wage rates are paid to compensate for the disagreeable labor conditions of item 4.
7. Progressive management accepts new cost-cutting technology.
8. Progressive labor leaders recognize the need for increased productivity.

The list of activities for which robots have been used in the automotive industry is extensive and is predicted to increase for pick and place assembly of small parts (less than 5 kg). The richness of the technology is attested to by the breadth of robotic activities [Engelberger 1980]:

1. spot welding: (1) framing, (2) respot, (3) body sides, (4) underbodies, (5) front structures, and (6) subassemblies;
2. pedestal welding;
3. arc welding: (1) stud welding, and (2) die casting, including extraction, die care, ladling, inserts and trimming;
4. permanent mold casting;
5. press transfer;
6. forging;
7. heat treatment;
8. casting cleanup: (1) shakeout, (2) gate burnoff, and (3) grinding flash; and
9. machine tool loading and unloading: (1) conveyor transfer, (2) palletizing, (3) plastic molding, (4) paint spraying, (5) adhesive layup, (6) foam layup, (7) glass handling, and (8) dimensional inspection.

The fundamental factor that will increase robot use in the automobile industry is the difference in cost between a robot and human labor [Engelberger 1980]. Including all fringe benefits and overhead expenses, the cost of an American automobile worker in 1980 was $15/hr. If the purchase price of a robot is $40,000 and the robot works two shifts for a useful life of 8 years the total cost would amount to $4.50 per hour as shown below:

- maintenance: $1.05/hr
- electric power: $0.35/hr
- cost of two overhauls: $0.40/hr
- depreciation: $1.25/hr
- interest on capital: $0.65/hr
- installation: $0.80/hr
- Total cost: $4.50/hr

In Figure 2, this hourly cost of robot use was compared to an hourly labor rate. The two curves diverge widely over time, confirming the cost advantages of robot technology.

One of the reasons why robots have been accepted in the automobile industry is their gradual introduction in a manner that displaces no current worker from his job. The employees have been reassigned to other duties and the robots have taken over the most disagreeable jobs.

The operating characteristics that make robots attractive to the automobile industry are:

- six infinitely controllable degrees of freedom (three in position, three in angle);
- fast programming that workers can easily learn;
- large memory size;
- random program selection;
- positioning repeatability to 0.3 mm;
- weight handling capability to 150 kg;
- intermixed point-to-point and path-following control;
- ability to follow moving objects on a conveyor belt;
- ability to receive directions from a higher-level computer; and
- high reliability, with at least 400 hr between failures.

Assembly of Small Objects

Current automotive industry use of robots is largely in arc welding and spray painting of auto bodies. A large increase in mechanical assembly is predicted by the GM Manufacturing Development Group.

In an analysis of the use of robots for the assembly of automotive systems, GM found that small robots with the ability to lift weights of 5.5 lb (2.5 kg) or less would be required for many functions presently done by humans [Beecher 1979]. A special robot that duplicated the dimensions and functions of the human arm and hand was designed. The system was called Programmable Universal Machine for Assembly (PUMA) and was designed to occupy the same floor space on an assembly line as a human. The machine works at a comparable speed and can be interchanged with humans for many different assembly tasks. Provision to use sight and touch sensors is included.

The first PUMA was delivered by Unimation, Inc., in May 1978, and has been used for a variety of tasks. Hot electric motor armatures are transferred by the PUMA to a commutator press, and then to a curing oven. Screws and rivets have been inserted into an assembly. Careful work has gone into solving the problem of parts orientation and pickup. As additional units are introduced into regular manufacturing service, the potential for using robots in light assembly operations will be determined.

Provision of the PUMA or other robot with a vision system has also been pursued by developers of advanced manufacturing systems [Dewar 1979]. The application that would be most common in the automobile industry is location of objects for pick and place assembly. Employment of the video camera and computer processing equipment normally used for machine vision would be too expensive for many of the applications of interest in automobile manufacturing. Therefore, the development of simple vision systems has been emphasized. A typical problem is the location of a part placed on a conveyor belt so that it can be picked up by the hand of the robot. The CONSIGHT lighting apparatus was developed that projects a thin beam of light onto the conveyor belt. When a part traveling on the belt intercepts the light beam, a solid state diode array detects the change in the light pattern and instructs the robot to pick up the part at the known location on the belt. Other variations allow one type of part to be picked out of several different objects and allow the orientation of the part to be determined. A vision device mounted on the robot arm allows holes to be located for inserting bolts.

Advances in robots for automobile assembly will see the inclusion of:

- vision and touch capability;
- coordinated multiple-hand assembly;
- computer-directed trajectories;
- smaller-sized, general-purpose hands;
- voice communication; and
- increased safety.

Farm Machinery

The manufacture of farm machinery is a segment of the automotive market that is characterized by short production runs and great diversity in the output of individual implements. More than 4700 combinations of options can be selected for a large tractor. [*Iron Age* 1980]. The manufacturing processes are similar to those of automobiles, but the numbers produced are considerably smaller. The problem of maintaining high productivity in assembly line and part production is shared from the largest to the smallest builders of farm machines.

To increase productivity, John Deere and Co. and other large manufacturers are moving from transfer lines to flexible machining systems, which are automated clusters of machines chosen to batch-produce a class of parts. Smaller batch sizes and the need to minimize in-process inventory require a faster response time, which is obtained from a machining center controlled by a central computer. John Deere has eight machining centers in operation with the numerically controlled machines operated 24 hr/day. Allis Chalmers expanded from 8 to 13 centers in 1974 and has an automated core molding line under computer control in its foundry. Automation is also found at Allis Chalmers in sheet metal forming and welding, with some computer numerical control.

Truck Assembly

In Lancashire, England, a new automated assembly line is in operation. Seven computers and 200 terminals allow a minimal crew to assemble trucks from 6000 types of components through 50 or so separate assembly stages [*New Scientist* 1980a-c].

Once issued by a person on the floor, instructions for a part pass through one of eight microprocessor-controlled cranes in the warehouse. The crane transfers the part to a collection point, where a trucklift driver takes it to the required position on the assembly line. After it is fully operational, the factory should produce 425 trucks per week with improved productivity and quality.

Automatic Transmission Testing

One of the largest, most comprehensive automated control systems found in a survey by Bailey and Pluhar [1979] of automated factories was in operation at the Ford Motor Company Transmission and Chassis Division in Livonia, Michigan. A hierarchical network of 40 Texas

Instruments computers consisting of 1 master, 10 intermediate-level and 37 satellite computers controlled the final testing of completed automatic transmissions, and component-level testing of both the main valve body and governor assemblies. The final test subsystem includes 20 individual microprocessor-controlled test stands. It drives a completed automatic transmission through a series of performance tests lasting 2.5 min. Test data from each satellite is transmitted to the intermediate-level computer and passed to the main computer for repair correlation, quality analysis and warranty records. Test results are also continuously displayed to the operator at the test stand. Defective transmissions go for tear-down and analysis to repair stations, where all test data are available to advise the types of corrective action needed. Terminals in supervisor's offices allow monitoring of the complete system and intervention. Two other terminals allow for programming development, improvement and testing.

Engine Testing

A process control system has been developed for production diesel engine testing [Ritter and Mundy 1971]. The process control system accepts analog and digital signals from transducers measuring revolutions per minute, torque, inlet air temperature, inlet and outlet fuel temperatures, main gallery oil temperature, piston cooling oil pressure, turbo burst pressure, fuel flow, and several safety parameters. Figure 6 shows the configuration of the process control computer.

The net results of using an automated system are increased test capacity (50% increase), regulated test cycle, data accuracy, safety testing and reliable historical data.

On-Board Microprocessors

The rapid advances being made in on-board microprocessors are likely to affect the entire automation field. Consequently, they are discussed here, although they are not directly related to automation of the automobile manufacture industry.

These microprocessors may solve the automotive industry's problems with meeting more stringent emissions standards. As the standards became more strict, exhaust gas recirculation and altered spark timing were found to be inadequate for the control of nitrogen oxides and too detrimental to fuel economy. The solution was the introduction of the three-way catalyst (TWC), which reduces nitrogen oxides, carbon monoxide and hydrocarbons. For this system to work properly, there must be

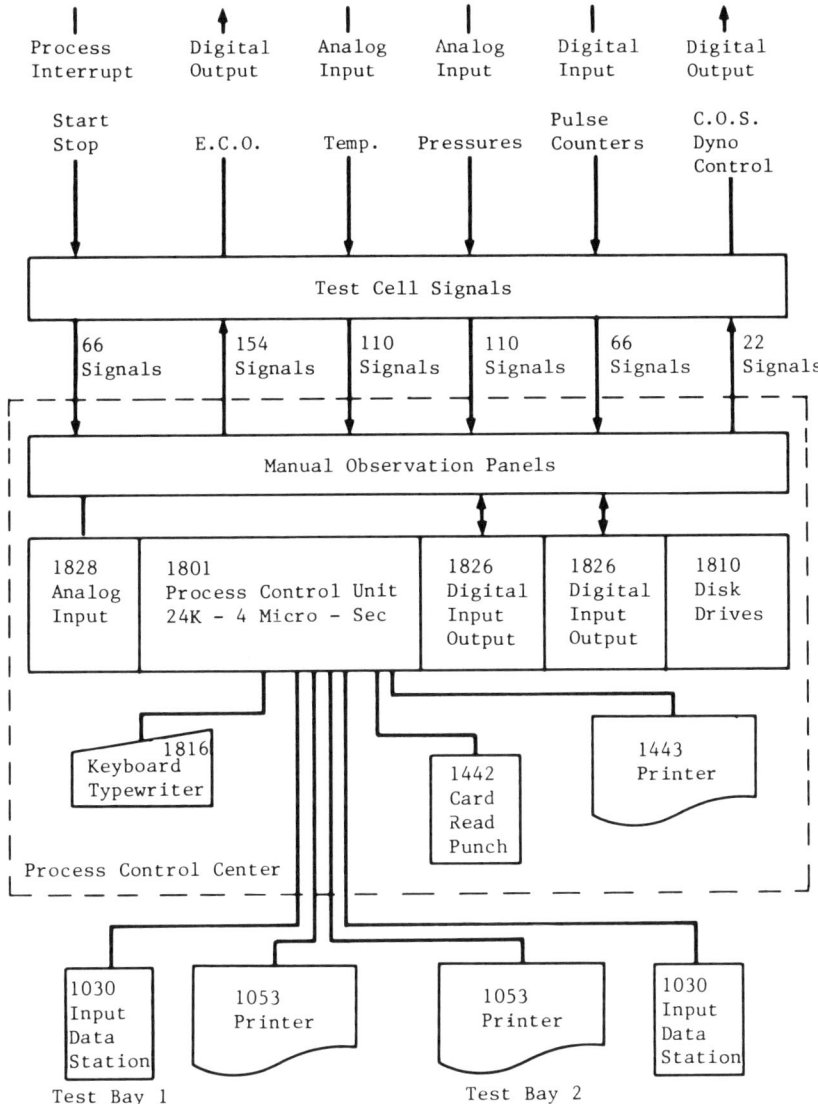

Figure 6.   Schematic of process control system [Ritter and Mundy 1971].

a lack of oxygen in the first stage and an excess in the second. The optimum method for achieving this is to operate just at the stoichiometric air-to-fuel ratio so all of the oxygen is used in combustion, but no more fuel than necessary is left unburned. Additional air can be injected into the second stage if necessary.

Thus, the complex task of maintaining a stoichiometric air-to-fuel ratio over a wide range of operating conditions is crucial to the effectiveness of the TWC system. In addition, exhaust gas recirculation and the need for sophisticated spark timing control increase the need for complex engine control. Add to this the necessity of achieving optimum fuel economy while meeting the emissions standards and it is clear that some new control system is required. The potential for that control system came from the realm of microelectronics [Lenard and Bluestein 1980].

The air-to-fuel mixture can be controlled by an electromechanical carburetor or fuel injection system to the stoichiometric point. This is generally referred to as feedback or closed-loop control. Although devices other than a microprocessor can provide this control, the microprocessor does it cheaply and efficiently; in addition, it can perform other tasks.

The use of microprocessors and semiconductor memories allows for such control through the use of multiple input systems with a complex look-up table of optimal spark-advance adjustments for a wide variety of engine conditions. Exhaust gas recirculation (EGR) can be controlled by look-up tables.

The block diagram for a generalized engine control system showing inputs and controlled functions is illustrated in Figure 7. Digital controllers in use today may have some or all of these features, depending on the level of control required for the particular engine. Although the need for these systems was first felt on larger engines, tightening standards have brought about their application to smaller engines manufactured by both foreign and U.S. automobile manufacturers.

Adaptive controllers are closed-loop systems with additional self-optimization capabilities. They can actively seek to optimize output by varying one or more inputs. The development of adaptive control of automobile engines is seen as the next major step in engine control. To date, controllers that can be called adaptive operate by continuously varying one engine variable to optimize the value of some measured engine value, often engine power. One of the systems now in production is a knock-sensing spark advance controller available on some turbocharged GM cars that incorporates some principles of adaptive control. The system uses an acoustic sensor to detect engine knock and then retards the spark by a preset amount. After a set time, the system gradually advances the spark to the original setting or until knock is again detected.

Two other examples of an adaptive control are a system that optimizes spark timing and an experimental system that optimizes the air-to-fuel ratio for fuels with varying alcohol content. Both examples measure

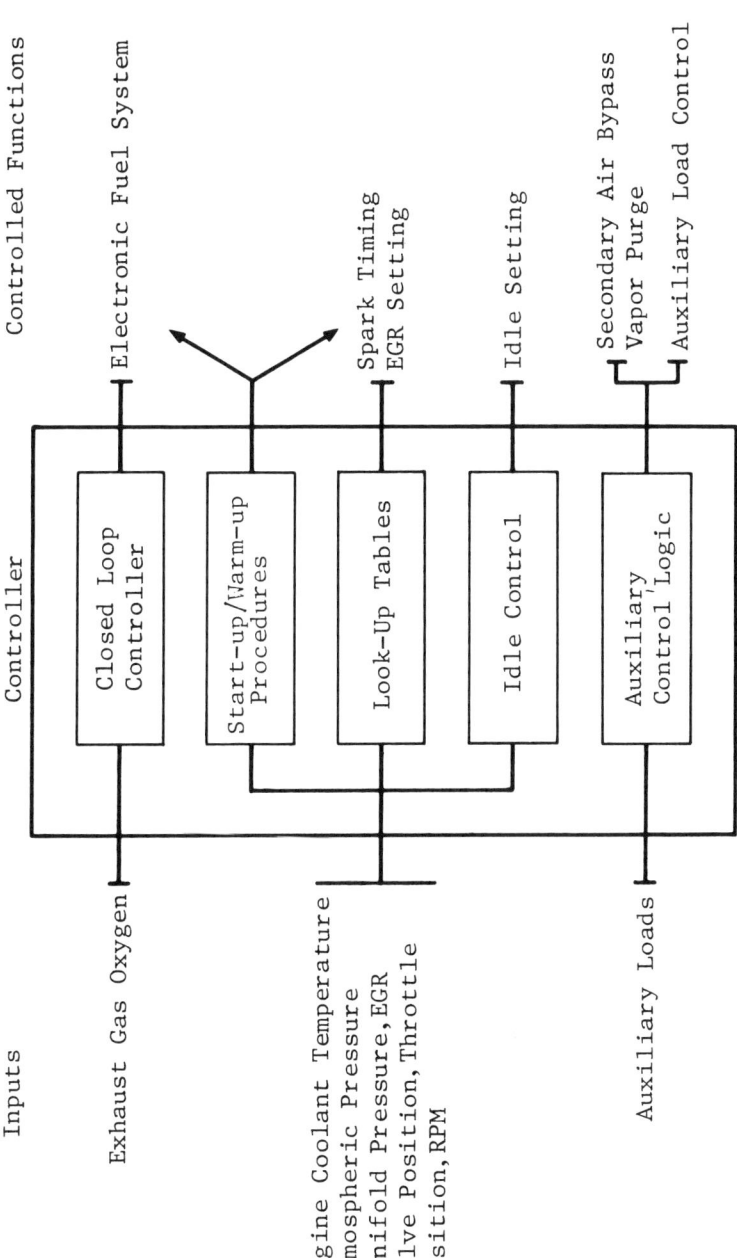

Figure 7.   Block diagram of generalized microprocessor engine controller [Lenard and Bluestein 1980].

engine performance by analyzing the power pulse of each combustion stroke as transmitted by the crankshaft motion. Such systems have fewer parts than equivalent preprogrammed control systems.

Will microprocessor control solve the problem of achieving high fuel economy while maintaining low emissions? As long as the TWC is a part of the system, the air-to-fuel ratio must be fixed at the stoichiometric level, leaving spark timing as the main determinant of engine efficiency. Although microprocessor control has produced fuel economy improvements of up to 10% in its first generation, the improvement is likely to quickly reach a plateau as optimization of the remaining factors is accomplished.

To reduce fuel consumption, ideas such as intermittent cutoff of several cylinders are being employed. This conjures up the possibility of an electronic engine where the microprocessor responds in the most efficient way, turning off all or part of the engine or radically changing spark timing and air-to-fuel ratio as needed. Such a system might use a number of microprocessors, distributed around the vehicle, connected to a common data base, and sharing monitoring and control functions.

The value of microprocessor control extends beyond the engine to functions such as optimized electronic transmission operation. Such vehicle systems would be valuable regardless of the type of engine used; moreover, microprocessor control could be usefully applied to almost any type of internal combustion, hybrid or advanced heat engine.

More in line with current perceptions of need in consumer-oriented items may be microprocessor-based safety applications. Antiskid and active collision-avoidance systems have been and are being developed. As in other cases, although digital electronics increase the practicality of these systems, their use will depend on other technical and social factors.

Existing systems provide the consumer with warning of burned-out bulbs as well as of the usual oil pressure and coolant temperature. However, a more important potential role lies in the diagnosis of malfunctions in the control system itself and in an expanding number of other engine functions. As control systems monitor more engine functions for control purposes, this information will also be monitored for diagnostic purposes.

AEROSPACE

Industry Characteristics

The growing aerospace industry employs 650,000 workers in 1274 establishments. The value of shipments of complete aerospace vehicles is expected to total $29.4 billion in 1980−22% above the 1979 value of $24.1 billion.

This increase reflects the steady rise in orders for large commercial transports that began during the first half of 1977. Shipments of large transports (31 seats and over) for 1979 are estimated at 381 units with a value of $8.1 billion−up 54% in units and 88% in value from 247 units valued at $4.3 billion in 1978. Shipments in 1980 are expected to reach 450 units with a value of $11.4 billion−up 18% in units and 41% in value over the estimated shipments in 1979.

General aviation (nonmilitary and nonairline) accounts for 96% of the Western world's civil aircraft, 80% of all civil flight hours and more than 50% of the world's passengers. Its 300,000 aircraft use less than 7% of total air-travel fuel. Of these aircraft, 75% are located in the United States, where in 1978 they used only 11.1% of aviation fuel, while transporting 112.5 million or 29% of intercity passengers.

Shipments by U.S. builders of general aviation aircraft in 1980 should total 16,100 units with a value of $4.3 billion−a 6% decline in units but an 18% increase in value over 1979. Table VI provides detailed breakdown of units produced and value of shipments from 1974 to 1977.

The aerospace industry is a high-technology industry where changes in processes, products and materials take place rapidly. These technological innovations (Table VII) are directed to labor reduction, cost reduction, and overall efficiency and product quality improvement.

Aircraft CAD/CAM

The technique of CAD drafting was extensively used by McDonnell Aircraft Company on the design of the F-18 aircraft, which was begun in 1975 [Zadarnowski 1980]. Systems using CAD included advanced composite materials, sheet metal parts, aircraft tubing, honeycomb structure and machine parts, and the entire horizontal stabilizer structural assembly. The percentages of parts produced using the automated system were:

- primary structure drawings: 21%
- primary structure started on the computer: 34%
- advanced composite parts: 58%
- sheet metal details: 50%

More than 1600 flat pattern drawings for composite-ply materials were produced. The large size of the ply flat patterns, the variations in complex patterns, the tolerance control required and the short time available

Table VI.   Shipment of Aerospace Vehicles and Equipment (value in millions of dollars) [DOC 1981]

	1974		1975		1976		1977	
	Units	Value	Units	Value	Units	Value	Units	Value
Aircraft and Aircraft Services, Total		10,502.2		11,223.0		11,427.0		12,047.2
Complete Aircraft, Total	16,970	8,357.0	16,825	9,017.3	18,017	8,943.3	19,685	9,325.3
Complete Military Aircraft	1,900	3,325.0	1,739	4,049.7	1,376	4,296.4	1,177	4,578.1
Complete Civilian Aircraft	15,070	5,032.0	15,086	4,967.6	16,641	4,646.8	18,508	4,738.2
Fixed-Wing, Total	14,261	4,877.5	14,248	4,701.9	15,820	4,323.4	17,524	4,354.4
Multi-engine	2,785	4,599.8	2,860	4,274.0	2,885	3,921.1	2,941	3,845.4
30-Place and Under	2,468	392.9	2,575	367.6	2,668	766.1	2,782	1,173.4
31-Place and Over	317	4,206.9	285	4,006.3	217	3,155.0	159	2,672.0
Single-Engine	11,476	277.7	11,388	328.0	12,935	402.3	14,583	508.9
Rotary-Wing, Total	809	154.5	838	265.6	821	323.5	984	351.7
Aircraft Services, Total		2,145.2		2,205.7		2,483.7		2,730.9
Modifications, Conversion and Overhaul		495.0		516.5		754.9		746.9
Other Aeronautical Services for Aircraft		1,650.2		1,689.2		1,728.8		1,984.0

Table VII.   Major Technological Changes in the Aircraft and Missiles Industry [DOL 1975]

Technology	Description	Diffusion
Advanced Filamentary Structural Companies	These new materials permit very large structural elements to be made in single pieces. Composites can be molded to exact dimensions, avoiding the high cost of machine tools required for machining metals. Their use also greatly reduces or eliminates the material wasted in scrap and can lower the weight of aircraft.	Recently introduced: use continues to expand.
Glass-Reinforced Plastic Rotors for Helicopters	These new rotors permit higher gross weights, altitudes, and airspeeds. Cost per pound should be comparable to metal rotors using automated production processes.	This is a new product and it has not yet had any impact on the industry.
Explosive Bonding	In this process the clad metal is placed parallel to and a slight distance above the backer metal. The clad is propelled across the standoff space by detonating an overlayer of explosive. A plastic flow of the metal surfaces slightly ahead of the collision point and causes a jet to form. This removes surface films, which can spoil the bond. Also, explosive bonding can eliminate the problem of brittle compound formation.	Already being used in the industry but the equipment is expensive and is only feasible for large production runs. The high cost is likely to hinder its rate of diffusion.
Bulge Forming Technique	This technique can cut manufacturing costs by 80%. It involves placing a cylindrical piece of annealed tubing in a die of the desired shape and then filling it with liquid or polyurethane. Advantages include elimination of joint strength problems and the need for welding.	Not in widespread use but its advantages may make it more popular.
Electrochemical Machining	Electrochemical machining is a quick way to manufacture large jet parts. Metal is removed from a workpiece by means of a formel electrode. It makes deep cuts quickly and eliminates the need for many basic operations.	Use should continue to expand.

Table VII,   continued

Technology	Description	Diffusion
Numerical Control	This process involves the automatic operation and control of machine tools by a system of electronic devices, servo-mechanisms and coded tape instructions. It reduces errors and machining time and can cut unit labor costs.	Already widely used in the industry.
Adaptive Control	A refinement of numerical control which allows a continuous automatic adjustment of the cutting process to compensate for such factors as vibration and temperature change. It increases machine productivity.	Adaptive control should become more widely diffused because of its advantages over numerical control.
Direct Numerical Control	This process uses a central processing unit to run the various machine tools. It has a number of advantages over numerical and adaptive control and users of these two may eventually switch to direct numerical control.	Direct numerical control equipment is already being used; use is likely to expand in the future.
Lasers	Laser technology has continued to expand and lasers are now being used for such things as trimming resistors, balancing gyrops, and welding.	Lasers are already widely used and their usage should continue to grow.
Computers for Simulation	Computers are being used to create models which permit the study of dynamic systems. These models can be used in evaluating situations which could arise under actual conditions.	The use of computers for simulation could expand as techniques are further developed.

to perform the design made this an ideal CAD application. The outlines for cutting the material with a numerically controlled laser were prepared along with bonded assembly models and production aids. The reduction from the number of drafting board hours required to computer terminal hours required was six to one. A cost analysis by McDonnell-Douglas showed that changes in procedures could more than double productivity.

Many sheet metal parts had to be pressed from rubber dies. Both the die and metal parts were produced numerically with an increase in productivity of four to one. Although original drawings were produced for this product as required by the government contract, it is expected that procedures will be changed so that the original will be recorded only on microfilm.

Use of CAD for tubing and electronic wire bundles eliminated the necessity to build a mockup for determining layout. Use of CAD for the horizontal stabilizer reduced design costs by 23% from the traditional design procedures. Due to the excellent part-to-part fit, production drawing changes 90 days after parts release, previously required to compensate for irregularities in dimension, were reduced 43% for small engineering changes and 70% for major drawing revisions. No data are available for the application of CAD to major assemblies such as the wing or the fuselage.

Preliminary Aircraft Design

A minicomputer-based system for performing the preliminary design analysis for military aircraft concepts has been developed by North American Aircraft. The configuration development program takes the initial aircraft requirements and sizing data and converts them into a geometric outline of the aircraft. The wings, tail, engine and fuselage are laid out on a cathode ray tube (CRT) display instead of a paper drawing. Electronic equipment, radar, weapons, landing gear, fuel tanks and other aircraft components are added and adjusted to meet weight and space requirements. The preliminary aircraft design is then released to analysis groups to calculate aerodynamic performance.

Numerous design iterations, ranging from minor variations to complete redesigns, are required to meet performance capability and cost goals. With manual redesign and drafting procedures, each iteration requires a new drawing that involves large expenditures of time and money. Since the preliminary designs are often performed and evaluated during the limited time available for submitting a proposal, reducing the time required to revise a design is very important. Programs running on mainframe computers were available to perform work of this type but were expensive. By implementing a program on a minicomputer, the cost was reduced to less than $10/hr.

In mid-1977, North American Aircraft developed its Configuration Development System, an interactive graphics design, analysis and drafting program. The system performs 80% of the work involved in drafting a preliminary design and can provide CRT views from the side, top, front or rear, and orthographic perspectives or cross-sectional perspectives at any station. To connect the surfaces defined by the designers, the program can calculate the points between defined surfaces and fill in the connecting surface.

The design begins by defining about 50 components, such as the fuselage, wing and engine dimensions. The system has a memory that stores the dimensions of more than 150 frequently used components such as pilot ejection seats, landing gear, wing profiles and weapons. A designer can create a new component in 15 minutes and can place it on the aircraft in several minutes. The computer will then perform calculations to determine the sizes of tires, struts and tail surfaces, aerodynamic drag, and cost estimates. Later the design is passed on to other computer programs for in-depth analysis of performance.

An example was given of use of the system for the design of a low-cost antitank aircraft. Examples were given of the graphics developed at each stage in the design. The example design required 6 hr of computer time and cost $36. The example would have required 30 hr of drafting time and another 20 hr each for aerodynamic drag and center of gravity calculations if performed manually. Both time and costs are reduced by 60%, with redesigns and tradeoff studies promising even greater savings.

Computer-Aided Graphics

The benefits of using computer-assisted graphics during the design stage of a military aircraft have been reviewed [Feder 1977]. The aircraft must fulfill complex and conflicting requirements that have no unique solution and must be evaluated in a coordinated interdisciplinary environment on a short time schedule. The ability to rapidly revise designs and evaluate suggested modifications is a key factor. The design process is an iterative one with numerous components of the aircraft affected by a change to one subsystem. The ability to visualize the effect of changes using interactive computer graphics is essential and allows numerous modifications to be evaluated in a much shorter time than if results were printed on paper in batch runs. The authors predict that the use of these systems will greatly increase as computer costs decrease.

After the design is completed, parts manufacturing begins, using the design information stored in the computer. Machining of metal parts and laying of composite materials is planned from the basic design data. Information such as tolerances, coefficients of expansion, tooling holes and reference surfaces come from one data base used by all departments. Quality assurance checking is performed by comparing the dimensions of the produced part to the original dimensions stored in the computer.

Another advantage can be obtained from using more complicated computer graphics systems that can display components in three dimensions. Traditionally, installation of electronic cables, fuel lines, subsystems and other components in mockups is required for placement. Using three-dimensional computer graphics, alternatives can be evaluated rapidly, and component placement can be observed from numerous angles, to eliminate interference without the need for constructing mockups.

Automated Aircraft Components Factory

The largest and most highly automated machining facility in the United States is the McDonnell-Douglas parts manufacturing plant in St. Louis, Missouri [King 1979]. Developed to fulfill the "zero defects" philosophy required for the manned space flight program, the use of computers for control of machine tools, production management and inventory control has greatly expanded with heavy government support. Parts for the F-15 fighter aircraft are being produced by almost 100 computer-controlled machines attended by a greatly reduced number of workers. A floor area of 24 ac is operated under a hierarchy of numerical controllers dominated by six mainframe computers.

The CAD/CAM process begins with parts design at an interactive graphics terminal and passes to production planning. Production tool design proceeds from the digital data base generated in the design. Parts are produced on a large number of numerically controlled machine tools whose operating programs were developed from parts design. The tools are operated under direct numerical control from the computers. Quality control inspection is performed on direct numerically controlled machines for comparison with the original design. Composite materials are cut and assembled on numerically controlled equipment; electronic circuitry is electrically checked on a computer controlled test wire analyzer. Metal tubing is automatically bent to the correct dimensions.

The management reporting system is a combination of data collection facilities, tabulation routines and report generating functions. Maintenance and production schedules and breakdown analyses are performed. A bill of material is assembled for each part produced and withdrawal and control of raw material are performed. The system is constantly modified and updated with new software routines, new computational equipment (particularly microprocessors) and new capability.

The benefits that McDonnell-Douglas claims have been obtained from producing numerous parts from an integrated CAD/CAM system are:

- increased productivity;
- better management control;
- greater design freedom;
- shorter lead time;
- greater operating flexibility;
- improved reliability;
- reduced maintenance; and
- reduced scrap and rework.

Examples of the benefits include the ability to redesign complex parts so that they may be machined more cheaply, ability to hold tolerances and reduce rejects (20% increased production), transfer of a job from one machine to another, and changing the program for a part without confusion or error. An analysis revealed a 55% improvement in scrap production rate when machines were connected to numerical control. Proper operation of a machine tool by numerical control reduces maintenance requirements. An automatic checking routine that stops the machine when an invalid instruction is transmitted prevented five parts valued at $80,000 from being scrapped. The cost of installing the checking routine was $20,000. The maximum metal cutting rates are possible with computer control.

One of the first products designed using this approach was a combined fuel and sensor housing for the F-15 fighter requiring 48 machine bulkheads. Conventional design and manufacturing methods would have required 11 months from project go-ahead to delivery of the first product. Using CAD/CAM the product was delivered in 5 months, at a substantially lower cost and without using blueprints. CAD/CAM is used also for the design of general aviation aircraft [Johannes 1980].

Integrated Program for Aerospace Design

The National Aeronautics and Space Administration (NASA) has sponsored a program to establish specifications and systems to incorporate computer aided design capability into aerospace procurement [Fulton 1980]. The NASA Integrated Program for Aerospace Design

(IPAD) will optimize productivity in the design stage. IPAD will complement the U.S. Air Force (USAF) Integrated Computer Aided Manufacturing (ICAM) program, which will optimize manufacturing productivity.

The Boeing Aircraft Company has been selected as the contractor to develop the IPAD system. An Industry Technical Advisory Board, consisting of computer and software suppliers and CAD users, provides guidance and direction. The methodology being developed is also applicable to projects besides aerospace, such as automotive, ship building and large civil engineering projects.

IPAD is a set of documents and computer software for implementing a project management system that has been found effective in aerospace projects. The interactions between design and manufacturing are specified for the definition, modification, change authorization and release of all documents and specifications required to carry out a large project. The software operates in a "user-friendly" language on a hierarchical, distributed, interactive network that is compatible with several existing computer systems. Standards are established for:

- engineering data base and documentation system;
- graphics and data base modification;
- establishing cost and schedules; and
- communications between computers in the network.

IPAD integrates design information, data bases, management of engineering information with computer graphics systems, design calculations, geometric modeling capability and other components of CAD into a single standardized approach that can be adopted throughout the industry. Interface problems between contractors and standardization within operating units of the same company will be improved with the establishment of a uniform technology for performing engineering design and management.

Although IPAD is being developed for use by the aerospace industry, application of the methodology to large civil engineering projects, ship building, automobile manufacturing and computing system development is also expected to occur.

Integrated Computer-Aided Manufacturing (ICAM)

The USAF [Honeywell 1980; USAF 1979,1980a,b], as the lead service for the Department of Defense, is running a large, long-term program to develop, demonstrate and use the most modern automated methods in the manufacture of weapons systems. Because of the high cost and complexity of modern military aircraft, the investment in establishing the automated system will be recovered in the savings that occur when the system is used for production. The ICAM program and development plan will produce systematically related modules in sheet metal, electronics and composite materials that can be combined into an automated manufacturing system with a consistent architecture, data base, design standards, management control system and all other required components.

The primary goal of ICAM is the production of compatible and standardized techniques that will shorten the time required for adopting automated manufacturing methods. Industry is heavily involved in the program. Activities include:

- designing an automated sheet metal production facility;
- establishing a CAM system for aircraft electronics;
- using a robot on production line operations for the F-16 fighter; and
- analyzing the human factors involved in unsuccessful attempts to automate manufacturing.

Computer Integrated Manufacturing International, Inc. (CAM-I)

CAM-I [Wisnosky 1979] is a nonprofit corporation seeking standardization by voluntary compliance from the dozens of companies that form its membership. The companies provide computer systems and software, automated machinery and components, and they include many companies that use automated production methods or will in the future. Current projects include:

- geometric modeling;
- factory planning and management;
- manufacturing process planning for fabrication of parts;
- defining manufacturing architecture;
- establishing standards in computer languages and software systems;
- educating of personnel in industry; and
- integrating CAD/CAM systems into a long-range plan.

ELECTRONIC COMPONENTS

Industry Characteristics

An estimated 4000 establishments employing 480,000 people manufacture electronic components in the United States. Electronic components industry shipments were expected to dip slightly in 1980 to $19.4 billion, down 4% from a high reached in 1979.

Nonetheless, the industry is expected to grow, with electronic components increasing by about 9.4% annually from 1979 to 1984. The largest part of this growth will be in semiconductors, whose average annual growth rate is expected to be 13.4% in constant dollars. By 1984, semiconductors will account for more than 60% of the value of all electronic components shipments.

Electronic components are the basic building blocks supporting a wide array of electronic equipment manufacturing facilities. Such equipment includes computers, industrial controls, communications and navigation systems, stereos, television sets, electronic games, calculators, watches and automotive electronic systems.

The continued development of both product and process technology is a key factor in the future growth of the electronic components industries. This is particularly true for the semiconductor manufacturers. Many of the new devices being developed, such as advanced microprocessors and large high-speed memories, will be coupled with new forms of light-emitting diode (LED) and liquid crystal displays (LCD) or even fiber optic systems, which will make possible many more new products, increasing the demand for other components.

Advances in manufacturing processes such as the electron beam and X-ray lithographic systems for semiconductor manufacturing are key factors in the very large-scale integrated (VLSI) circuits program. This technological progress in the density of integrated circuits is being pursued vigorously by U.S. semiconductor manufacturers.

Automated Design

The fabrication of solid state electronic circuits and their assembly into functioning subassemblies is one of the most highly automated industries. Many semiconductor fabrication operations were automated because they must be performed under clean conditions and would be contaminated by human contact. The large numbers of items produced, numerous steps in the manufacturing process, the requirement for extensive quality control testing and the recording of a large number of performance parameters have also contributed to the adoption of automation.

The use of automated methods for the design, layout and assembly of components in the electronics industry is widespread. Many of the procedures are used because of the special requirements of fabricating the small dimension components that require a high degree of cleanliness. Electronic components are also significantly smaller and lighter than most components used in the assembly of discrete manufactured items. Although there has been little transfer of this technology to other industries to date, MITRE has found from personal communication with executives in the electronics field that at least two of the large computer companies are planning to market automated assembly equipment originally developed to assemble equipment and components in their own plants.

Table VIII presents a general automated semiconductor design cycle [Beardsley 1971; Moore 1979; Schensstrom and Williams 1976]. The physical placement of electronic components on printed circuit boards and the design of the interconnecting electrical leads for high-reliability aerospace electronics has been performed by a CAD/CAM system [Gargione 1980] at RCA Astro Electronics. The system has been in operation for four years and has been applied to more than 200 board designs. The system took over what had been a tedious job of manually produced drawings, schematics and numerically controlled inputs for the design of welded wire boards. Such welded circuits are important for two reasons: (1) short production runs are quickly prepared; and (2) wiring changes and circuit modifications can be made without affecting the components on the board. However, the designs are very complicated and demanding to work on manually.

The computer-aided process begins by producing a schematic drawing showing the components and their interconnections. Use of computer files showing the dimensions of components and connection requirements is used in making a list of pin connections. A numerical control

Table VIII. Semiconductor Design Cycle

Step	Description
Develop System Specifications and Logic	Customer input
Logic Simulation	Verification of logic design before hardware implementation (topological analysis)
System Partitioning	Division of equipment logic into a number of chips
Preparation of Logic Diagram with Cells	Description of circuit elements stored on disk
Equation	Equations define the logic throughout a particular chip
Logic Simulation	Chip design verified through logic simulation (circuit analysis)
Display	Artwork generation on CRT through the use of routine program defining cells and interconnections
Drafting	Lay out drawing on plotter

paper tape is then prepared for drilling holes in the board on which the components are mounted. The hole sizes and sequence in which the holes are drilled are determined by the computer. A paper tape is next prepared for the machine that stretches wire between the connections and welds it to the pins on the components.

The computer also prepares tabulations of parameters, including solder joint counts that are required for later steps such as modifications and quality control. Violations of design rules are found and signaled for correction.

Manufacture of the boards is performed with some manual effort, but mostly with numerically controlled machines that perform the operations and also check the quality of the work. The work of the designer is reduced without reducing his control over the process. Numerically controlled equipment reduces manufacture time, errors and scrap. Errors are so infrequent that electrical continuity tests do not have to be performed on the board as was required when the welding was done by hand. This method of assembly has reduced costs by more than 50%.

Large-scale integrated circuits can be designed, built and tested in three months for $35,000 with computer-aided design and design-automation (CADDA) facilities. Such automated development of circuits in a single cost-effective cycle may make low-volume custom large-scale integration (LSI) functions economically feasible.

The U.S. Army Electronics Research and Development Command has developed a complete CADDA system applicable to metal oxide systems and other advanced technologies. Its rapid design/fabricate/test cycle is based on the use of computers for logic simulation, area-efficient layout of box outlines that describe the size and input-output connections of standard cells, and generation of test patterns. A typical one-chip circuit containing more than 2000 devices produced by this program at a cost of less than $400, is smaller, cheaper, and more efficient than the integrated circuit (IC) equivalent.

Key elements of the CADDA system have been widely disseminated to defense systems contractors and are being applied to major pieces of U.S. Army equipment. The system is being enhanced to extend the availability of custom LSI to higher frequencies and greater degrees of complexity.

Some of the benefits associated with automation include [Groner 1981]:

- larger output of IC;
- higher accuracy;
- easier documentation;

- computational capability; and
- improved quality control.

Automated Semiconductor Fabrication

Semiconductor fabrication starting from a silicon wafer is a well established technology. The sequence of steps involved is shown in Figure 8.

An automated on-line processing system for wafer fabrication and testing has been developing over several years. The subject has been reviewed in a recent Institution of Electrical and Electronics Engineers (IEEE) publication [IEEE 1981]. Today most semiconductor manufac-

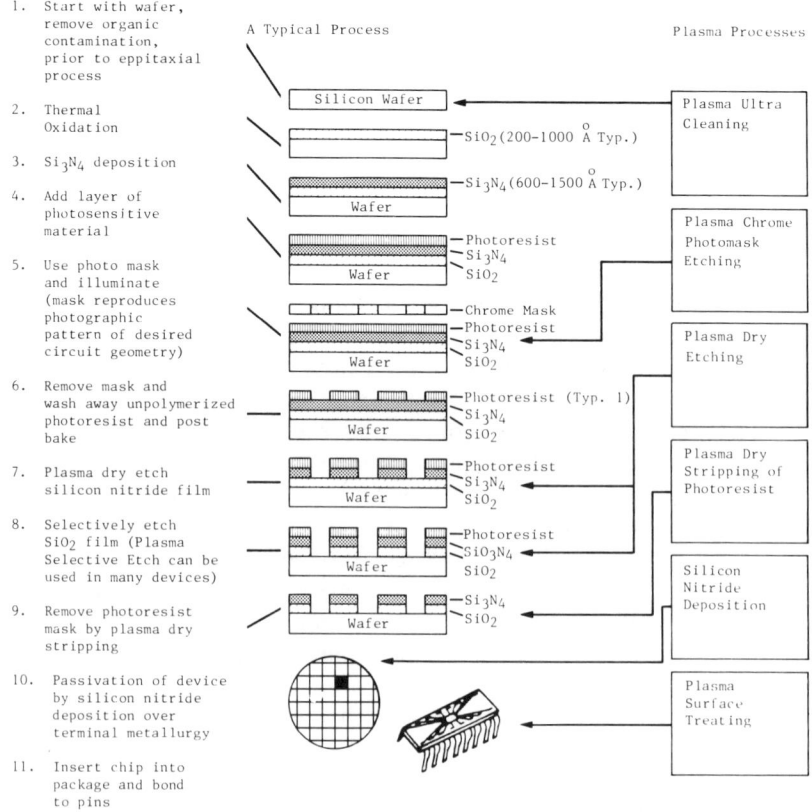

Figure 8.   Microcircuit fabrication [LFE Corporation).

turers are using some type of automated system [Malkiel 1978]. An automated system at the RCA facility at Palm Beach Gardens, Florida, reduced defects by 75%. Totally automated wafer processing is technically feasible, but production is usually done in a segmented fashion. A typical system might have each module driven by a microprocessor, with a communication system linking the microprocessors.

Semiconductor manufacturing is a repetitive process; however, wafers rarely go through the same process each time. Automatic on-line photoresist processing, automatic projection aligner, deposition system, automatic loader, wafer scrubber, baking, coating and transport systems are available. Automation is likely to continue in the industry toward the goal of improved device quality and final yield.

METALS

Industry Characteristics

The primary metal industry provides major input to the manufacturing segment of the U.S. industrial economy. Steel, ferrous castings, aluminum, copper, lead, zinc and tin represent 82% of the value added by all primary metal industries. Computers and automation are used in all of these industries, and this section focuses on the steel industry as the primary example.

Since this section reviews the use of computers in a variety of steel processes, it is useful to depict a "typical steel plant," or, more accurately, to identify the possible pathways from iron-bearing raw materials and steel scrap to a finished steel product. This is illustrated in Figure 9.

In 1977 the domestic steel industry used 3×10^{15} Btu (3.16×10^{18} J), or approximately 4% of the total energy consumed domestically. Figure 10 presents an estimate of this energy consumption by conventional unit processes. Blast furnaces are the largest users of energy, with 40%. Heating and annealing furnaces follow, with 15–20%. Coke oven, steelmaking furnaces and casting/breakdown operations each contribute roughly 10% of energy use. The remainder is used by other processes.

Between 1972 and 1978 the steel industry was responsible for 4.6% of U.S. energy consumption. Over this time period, the percentage supplied by coal decreased, while the consumption of petroleum and purchased electricity increased. Coal became more expensive and more difficult to obtain while prices for petroleum and electricity climbed.

The opportunities for increased energy-efficiency in the steel industry have been studied extensively. A 35% savings in total energy consump-

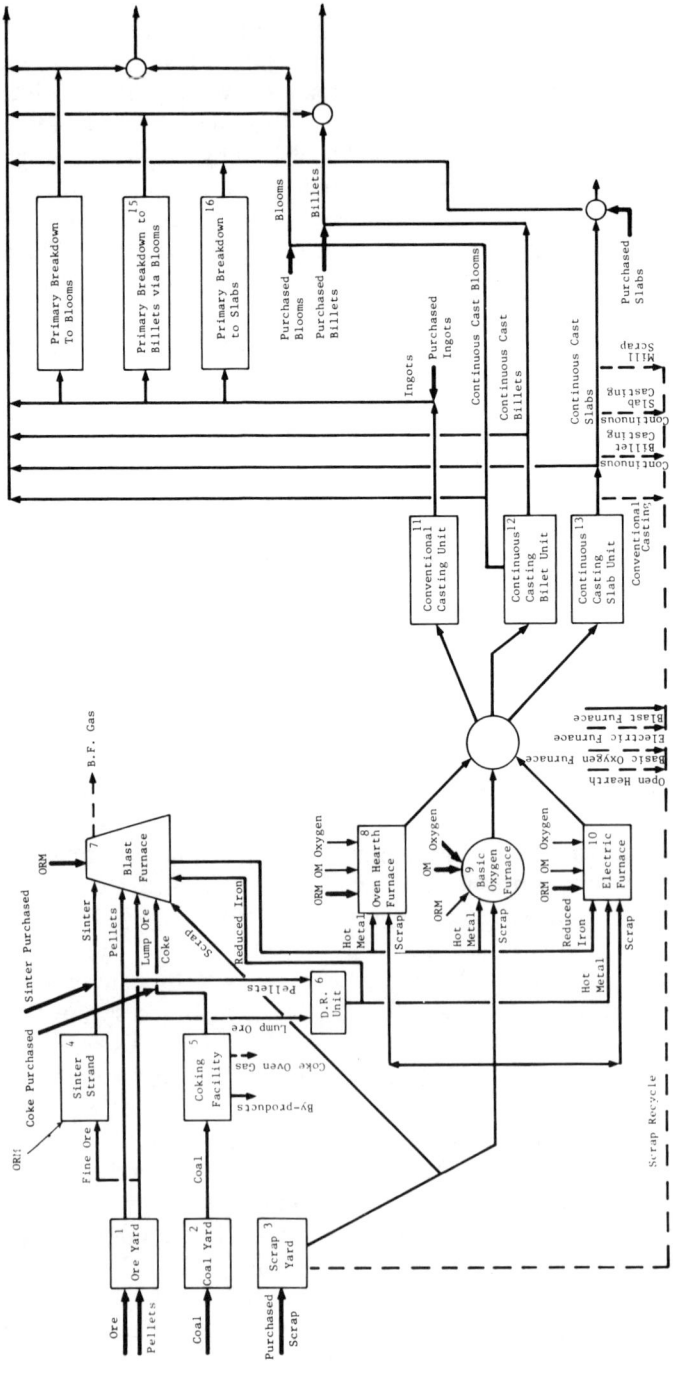

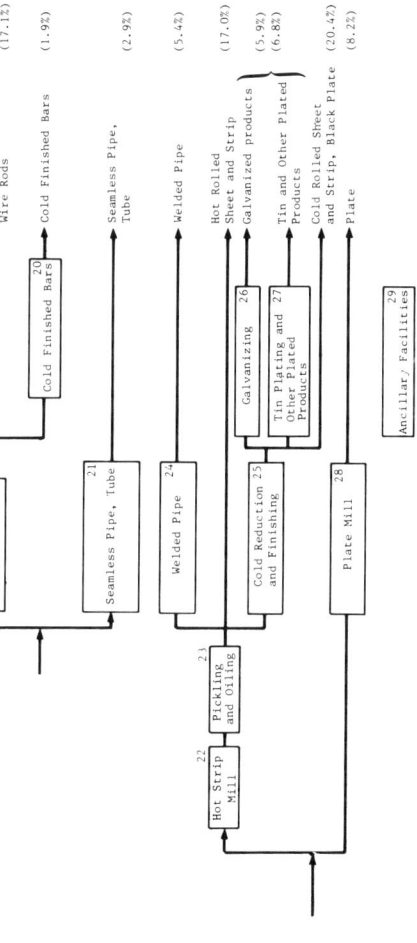

Figure 9. Process unit interrelationships. ORM = other raw materials, including items such as fluxes. OM = other metallics, including items such as ferroalloy additions. Each box, although representing a process unit, may include several steps. For example, coke facility includes coke ovens and by-products facility. At the end of the figure appear the different categories of carbon steel products shipped by the U.S. iron and steel industry. The figures in parentheses show the contribution of that category of shipment as a percentage of the total in 9712. Process units 11, 12 and 13 would include vacuum degassing operations when found in a plant. ――――― = in-plant flows; ------- = scrap recycle; ▬▬▬ = flows from outside; 0 = flow streams joined to obtain common cost to following process unit.

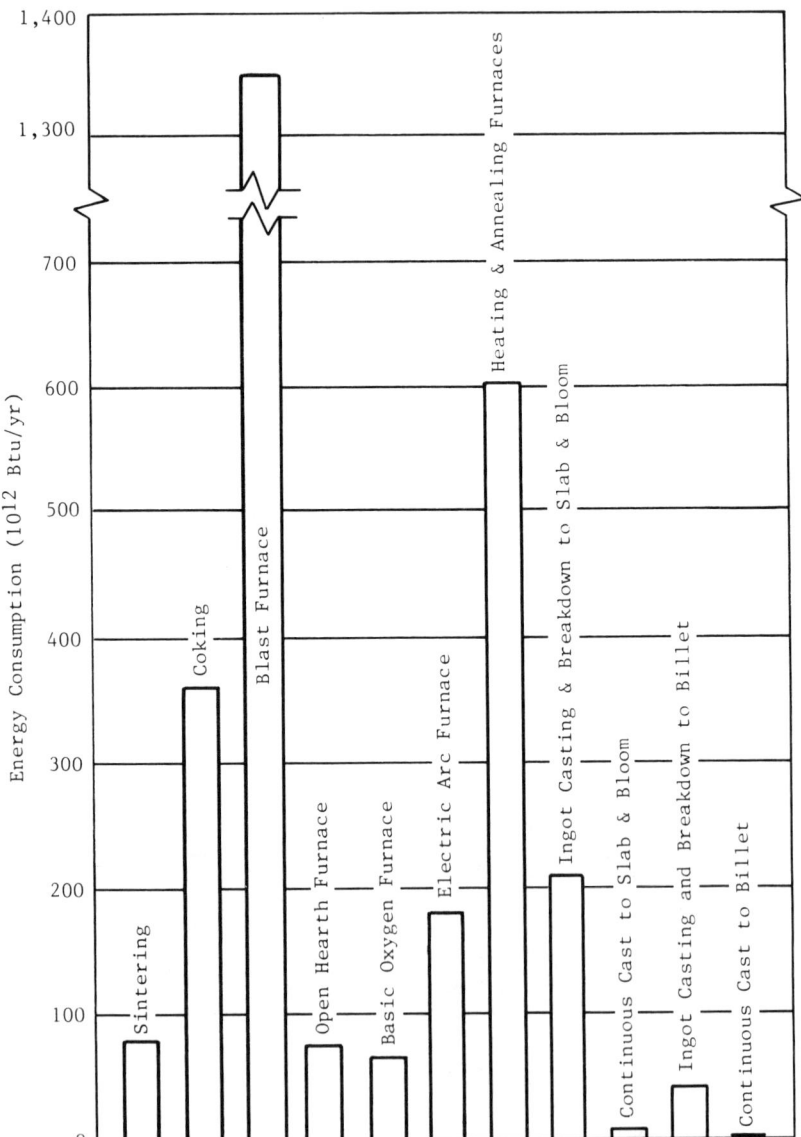

Figure 10.   Estimated U.S. energy consumption by major process units in 1977 [ADL 1979].

tion per ton of product shipped would be achieved if the oldest plants were shut down and replaced with modern facilities. However, the capital formation problems of the steel industry make this unlikely. Improvements giving marginal increases in the energy-efficiency of existing plants cannot compete for the limited capital available with the requirements for replacement of existing equipment and environmental control equipment. Other factors influencing energy-efficiency include:

- the ability to operate a plant efficiently at high capacity if the market demand for steel holds;
- the rising cost of environmental controls, requiring 2% of total energy consumption in 1975, and predicted to rise to 7% in 1983;
- the declining quality of metallurgical coal;
- the complex effects resulting from implementation of the National Energy Act (NEA) that determine the fuels available to the steel industry.

Future energy trends in the steel industry are uncertain due to an inability to predict the role that the federal government will play. Tax law revisions are being discussed that would encourage the construction of modern equipment and processes, resulting in energy-efficient facilities. Promotion of coal technologies and development of processes that produce metallurgical-quality coke from low-grade coal would help. Revision of environmental regulations would reduce the associated energy requirements and free capital for more energy-efficient facilities. The regulations formulated by the U.S. Department of Energy (DOE) to implement the NEA will influence energy decisions made in the steel industry.

The opportunities for energy conservation in the American steel industry [AISI 1978] have been assessed by several organizations and summarized for the industry in five basic processes and two other areas:

1. ironmaking: iron ore sinter plants, coke conversion and blast furnace;
2. steelmaking: basic oxygen process, open hearth furnace and electric furnace;
3. primary conversion: final product casting, ingot casting and continuous product casting;
4. hot working: forging, extruding and rolling;
5. finishing: annealing, heat treating and numerous other processes;
6. other energy-saving operations: material handling, general plant utilities and space heating; and
7. electric power generation in the steel mill: steam turbines and electric generators.

Process Control Trends

In a review of process control trends in the American steel industry, Monteith [1978] concluded that the use of the digital computer has advanced significantly since its introduction in 1956. Fewer than 25 installations used computerized control systems in 1965, but use of computers increased rapidly until today, when 300 installations can be found. Applications have been found in the areas of:

- raw materials;
- ironmaking;
- steelmaking;
- continuous casting;
- soaking pits;
- hot metal working;
- cold metal working;
- annealing and heat treating;
- finishing lines;
- plant scheduling and inventory;
- energy management; and
- chemical and metallurgical laboratories.

The application of process control technology had the following objectives in order of importance at the mills that were studied:

	% of Installations
increased productivity	65
increased yield	50
reduce costs	60
increase product uniformity	50
improve product quality	55
improve working conditions	8

In the late 1960s and early 1970s capacity expansion and modernization were the motivations for adopting computerized process control. Currently, cost and the availability of raw materials and energy are of greater importance. As business competition increases in the steel industry, improvement in product quality will become an important consideration. In times of overcapacity and limited capital availability, the ability of improved process control to modernize existing facilities by increasing productivity and reducing costs is important. Of the six objectives, increased productivity and cost reduction have historically been the most important.

Management will often place a steel plant under computer control with three expectations:

1. reduction in labor costs;
2. improvement in operating efficiency; and
3. optimization in complex conditions.

In practice, labor costs do not significantly decrease because low-priced labor is replaced by expensive labor for programming and maintenance. It is often discovered that the process has not been designed for computer control and that bottlenecks appear when the plant is not operating at design conditions. The second objective of improved operating efficiency is achieved after bottlenecks have been overcome and the plant has been equipped with adequate and reliable instrumentation. One of the major contributions of the human operators is that they are adaptable to unexpected situations. Computer systems are not programmed to be versatile. The ability to efficiently operate over a range of high and low capacity conditions is best achieved by developing a hierarchical computer control system in which a higher level system controls the operation of plant functions. To achieve this, middle management must be educated to appreciate the capabilities and limitations of computer techniques and methods.

Data on installed computers dedicated to specific process steps have been collected by Long [1978] (Table IX). He found 8 aimed at heat treating controllers, 11 in zinc coating, 2 in tinning, 4 in temper mills and 4 in auxiliary finishing roles. A total of 22 machines were dedicated to energy management in 20 installations; 31 computers were found in 27

Table IX.   Process Step Computer Use in the Steel Industry
[Long 1978]

Process	Number of Installations Considered	Number of Computer Installations Reported	Percentage Computer Installation
Blast Furnaces	141	20	14
Steelmaking Shops	81	52	64
Continuous Caster	25	8	32
Ingot Processing, Soaking Pit Monitoring and Control	80	2	2.5
Slabbing/Blooming Mills	81	8	9.9
Hot Strip Mills	34	18	52.9
Structural Mills	134	10/12[a]	7.5/9.0
Cold Mills	101	17	16.8

[a]Control/auxiliary.

chemical laboratories; only 3 performed in-plant scheduling, and 9 appeared in diverse functions such as stacker crane control and water utility management.

These machines may be single- or multifunctional, and they may operate in a variety of modes from simple monitoring and alarm to closed-loop control. Table X summarizes the functional use of computers in the industry sample surveyed by Long [1978]. While 47% of the surveyed installations employed process analysis instrumentation, only 20% used dynamic process control. Predictive models are used frequently in hot and cold strip mills in steelmaking and strand casting. Among the large steel companies, Long [1978] mentions three outstanding predictive models:

1. a fully automated basic oxygen furnace (BOF) end-point control based on experience at Bethlehem Steel Company's plant in Bethlehem, Pennsylvania;
2. a single stack coil annealing control based on a deterministic heat-transfer model at Armco, Inc., in Middletown, Ohio; and
3. a tandem mill closed-loop, short- and long-term adaptive model at Inland Steel.

Table XI lists the present-day applications of computers in the steel industry.

Compared to other industrial sectors, the iron and steel industry has

Table X.   Mode of Use of Installed Computers[a] [Long 1978]

	%
Single-Function Machines	
Monitoring and alarm only	6.0
Open-loop advisory control	7.5
Closed-loop supervisory control	11.0
Closed-loop DDC	10.0
Total	34.5
Single- and Multifunction Machine Features	
(percent of total)[b]	
Monitoring and alarm	58
Open-loop advisory	44
Closed-loop supervisory	47
Closed-loop DDC	44

[a]85% of the machines have closed-loop control.
[b]Several features may be found on a single machine.

Table XI.   Controlled Functions [Long 1978]

Area	Functions
Raw Materials	
Mining	
Crusher/concentrator	Monitor/log; alarm; maintenance schedule
Agglomeration	Grate bed depth control; closed-loop control; various functions
Traffic	CTC
Ore bedding	Conveyor operation; stacker/reclaimer
Sintering	Bin hopper level; blend; permeability; burnthrough; BF burden calculation
Coking (Ovens) and By-products	Coal classifying, treatment, blend; oven firing and temperature, push schedules; data logging; DDC, process loops
Ironmaking	Burden calculation; charge sequence; coke weight and moisture analysis; top gas analysis; fuel injection; blast moisture; cooling water; tuyere level control; data logging
Steelmaking (BO, Q-BO, AO)	Charge calculation; hot metal and scrap weighing; flux weighing and control; lance positioning; dynamic C/T measurement; dynamic turndown control; temperature prediction and control; gas flows and ratios; alloy additions; data logging; heat log production; ingot tracking; spectrometer control
Electric Arc Furnace	Charge calculation (least cost); flux-scrap weighing; operator guidance; power demand scheduling; alloy additions, Fce. and ladle; material inventory and control; monitor and logging; oxygen flow control; fume system monitor; log production
Open Hearth	Alloy additions; firing practice
Strand Casting	Logging; mold level; strand speed; spray water; tracking/inventory
Ingot Processing	Tracking; charge and firing strategy; fuel optimization pit firing control; pit mapping
Hot Working[a] Primary[a]	Reduced scheduling calculations; ingot tracking/pit map; scale removal
HSM[a]	Product tracking/reheat Fce. map/firing; mill pacing/dropout sequence; mill setup; AGC; crop shear; logging reporting; demand control; on-line equipment diagnosis
Plate/Structure/Bar[a]	Reduced scheduling; AGC; product tracking/scheduling/inventory; saw length control (hot and cold)
Cold Working[a]	Tread control (accelerate and decelerate); temper mill setup, tracking, scheduling and control
Finishing	
Coating	Line setup; line-speed; coating weight control; logging
Annealing, Box and Continuous	Fce. firing DDC; optimal box charge/scheduling; static soak time prediction; dynamic soak time control

Table XI, continued

Area	Functions
Coal Buildup	Logging; inspection
Heat Treat, Pipe,	
Plate, Bar	Product tracking; product temperature continuous (Aus. and temperature)
Schedule and Inventory	Product tracking; mill lineups
Support Functions	
Energy management	Demand control and scheduling; gas use and distribution; powerhouse control; data logging
Lab automation	Spectrometer control/analysis/reporting; met lab data handling
Environmental	Water quality (monitor/alarm)
Other Functions	Water utility management (control); crane control; roll quality (measurement)

[a]In many mills, hot and cold, the complete mill operation is controlled.

implemented computer controls in a large number of their operations. In most cases, energy management was not the primary reason for automation, although some energy savings have been realized, as indicated in Table XII.

Cokemaking

Three major subsystems of the cokemaking process have been automated: material handling, by-products and combustion control. Virtually all aspects of material handling have been computerized, including movement of raw and blended coal, coke, and quench gas, and oven reversal. By-products plant controls include the handling of ammonia, sulfuric acid, light oils and tar manufacture. An example is the Armco system, with 50 direct digital control loops, including 19 primary controls of level, 4 flow, 7 temperature, 2 pressure, 2 acid concentration, 2 cascaded controls of level, 2 temperature, 2 pH and 1 oxygen. An example of a combustion control is the system developed by Nippon Kokan KK (NKK) (Figure 11). The system relies on a series of thermocouples to provide control of the coke oven operations to:

- optimize combustion control;
- determine the end point of carbonization; and
- control the oven operation schedule.

Table XII.   Summary of Estimated Energy Savings by Application of State-of-the-Art Computer Control [DOE 1979]

Steel Mill Sector	Product	1977 Production (million tons)	Energy Use Per Ton of Product (million Btu)	Basis for Further Estimated Energy Conservation		Estimated Potential Energy Savings (10^{12} Btu/yr)
				State-of-Art Technology	Energy Savings	
Coke Ovens	Coke	52.7	3.5	NKK	3% or more saving	5–50
Blast Furnaces	Hot metal	81.3	16.6	Algoma, Outokumpu	1–10%	5–500
BOF	Raw steel	77.4	0.52	Kawasaki	Yield improvement up to 0.5%	5–50
Ingot Casting	Ingots	110	—	Kemlo '79	0.15–0.75% yield improvement	0.5–50
Electric Furnaces	Raw Steel	27.9	5.88	Mostly automation	Limited further improvements	
Continuous Casting (Computer control)	Slab, bloom, billet	15	0.67	Kawasaki	30% savings	0.5–5
Soaking Pits	Ingots	110	1.5		10% (ADL estimate)	5–50
Reheat Furnace	Reheated raw steel products	125.3	2.78	Lackenby (BSC)	5–25% reduction in fuel	5–50
Plate Mills	Plate	7.5	0.95	Kawasaki	0.75–2% increase in yield	0.5–5
Hot Strip Mills	Strip, sheet	56.5	1.05	Armco	2.5% increase in yield	5–50
Cold Strip Mills	Cold rolled strip, sheet	21	0.84	U.S.S. Fairfield	0.8–1.14% increase in yield	0.5–5
Annealing	Annealed strip, sheet, wire	21	1.64	Armco	10% fuel savings	0.5–5
Centralized Energy Management		—	—	NKK	Up to 5–6% better by-product fuel use	5–50

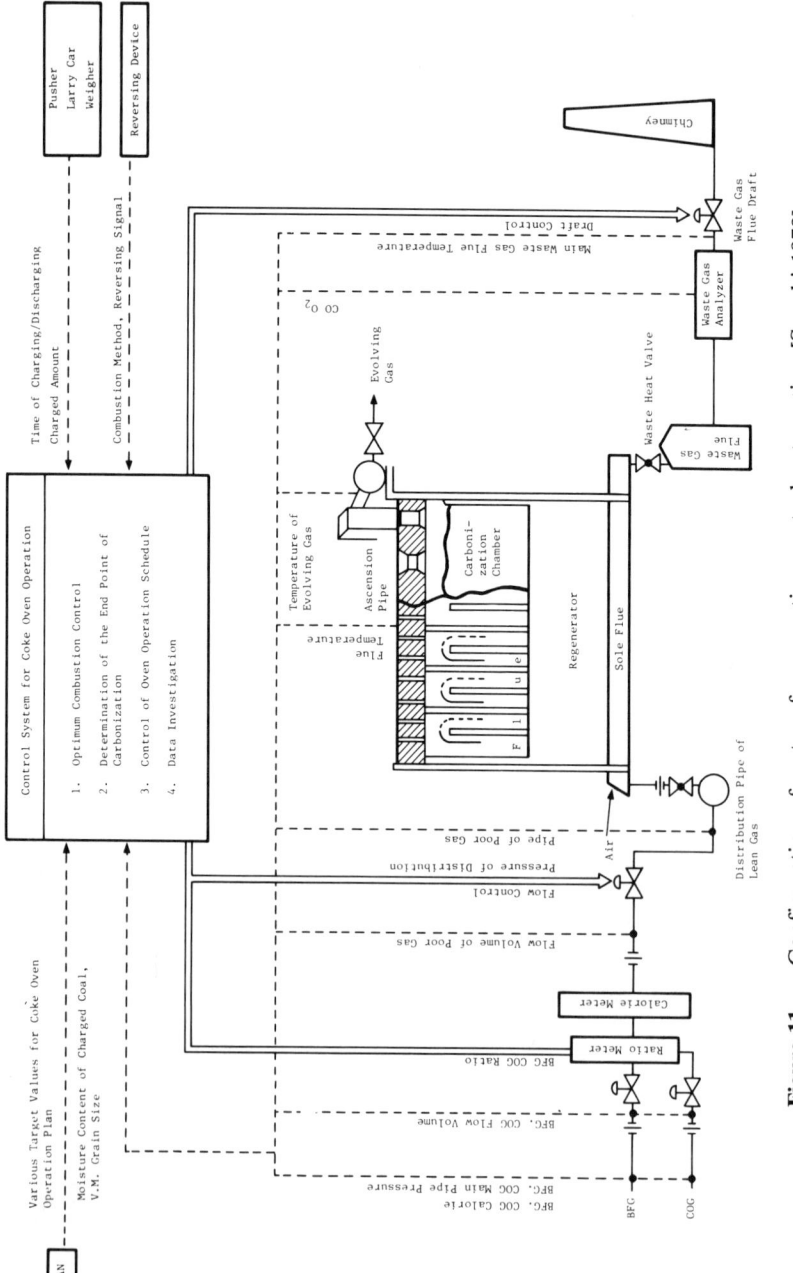

Figure 11.   Configuration of system for operation control automation [Suzuki 1978].

Benefits claimed for the above described system are [Suzuki et al. 1978]:

- improvement of coke quality ($+0.25$ coke strength);
- stabilization of coke quality (30% net coking time dispersion);
- savings in heavy coking coal (3%);
- improved productivity (3%); and
- reduction of heat consumption $-36,000$ Btu/ton of coal.

In operations conducted at the South African Iron and Steel Industrial Corp. [Farquharson 1979; Farquharson et al. 1978], the use of a microprocessor to control coke weighing and moisture control has demonstrated a reliability in a hostile environment greater than the reliability of the instruments and sensors that are being controlled. The ability to determine more accurately the quantity of dry coke fed to the blast furnace in relation to the other constituents of the charge results in more economical furnace operations and better control of the iron chemistry.

A commercially supplied process controller was built around a microprocessor having 16K and 8-bit memory and was used to control coke weighing. A nuclear moisture gauge permits the dry weight of the coke to be determined for every hopper. A sequence of automatic checks and alarms is employed and the weights are automatically logged (Figures 12 and 13). An improvement in the chemistry of the iron that has been produced by the use of the microprocessor is the reduction in variation of the silicon content. Before the computer system was installed, the standard deviation of the silicon content was 0.3%. After installation, the standard deviation was reduced to 0.2%.

Sintering

The material handling aspects of the sinter process involving bin hopper levels and ore blending have been placed under computer control. In addition, U.S. facilities have instituted control of permeability and burn-through.

Closed-loop control of the sinter plant is reportedly used at ARBED in Luxembourg [Funck 1978]. It measures iron content, level of return, fines, mechanical strength and sinter size, and provides control of fuel rate, moisture, return fines rate and bed height. Control of bivalent iron to $\pm 0.5\%$ and strength to a standard deviation of 1% have been reported. Energy savings attributed directly to the sintering process are small; however, quality control provides an opportunity for additional savings in the blast furnace.

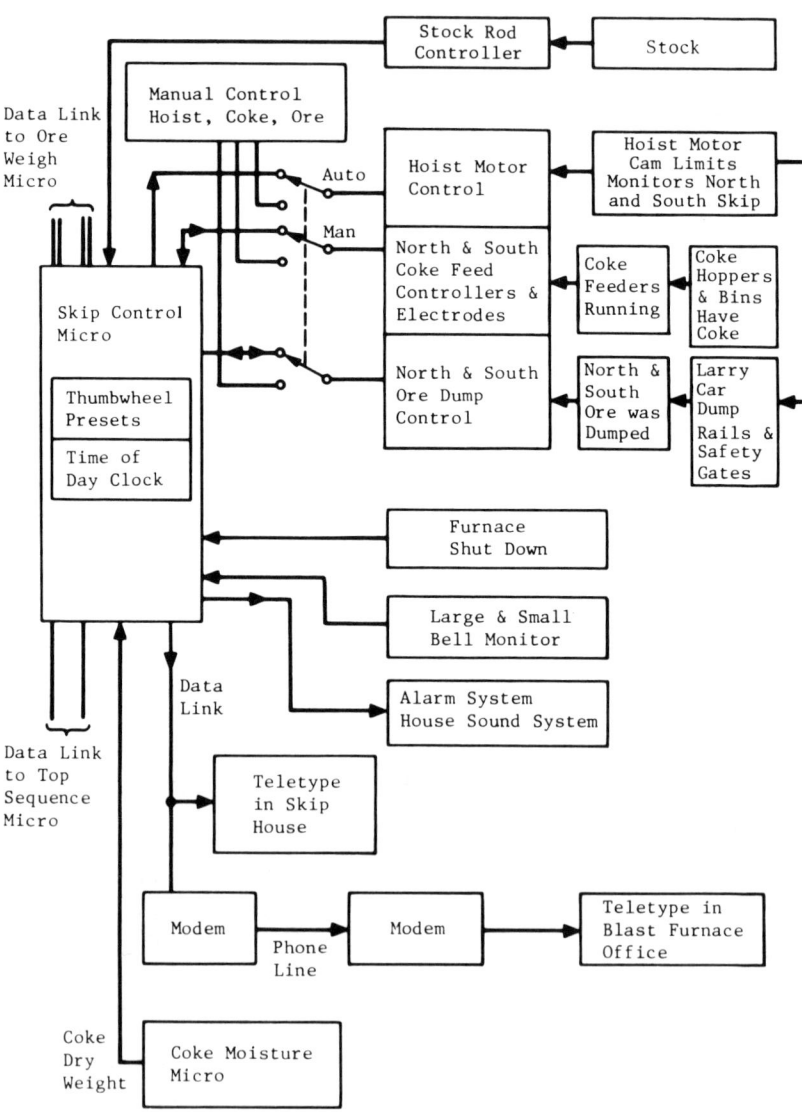

Figure 12. Blast furnace "J" skip control microcomputer system [Farquharson et al. 1978].

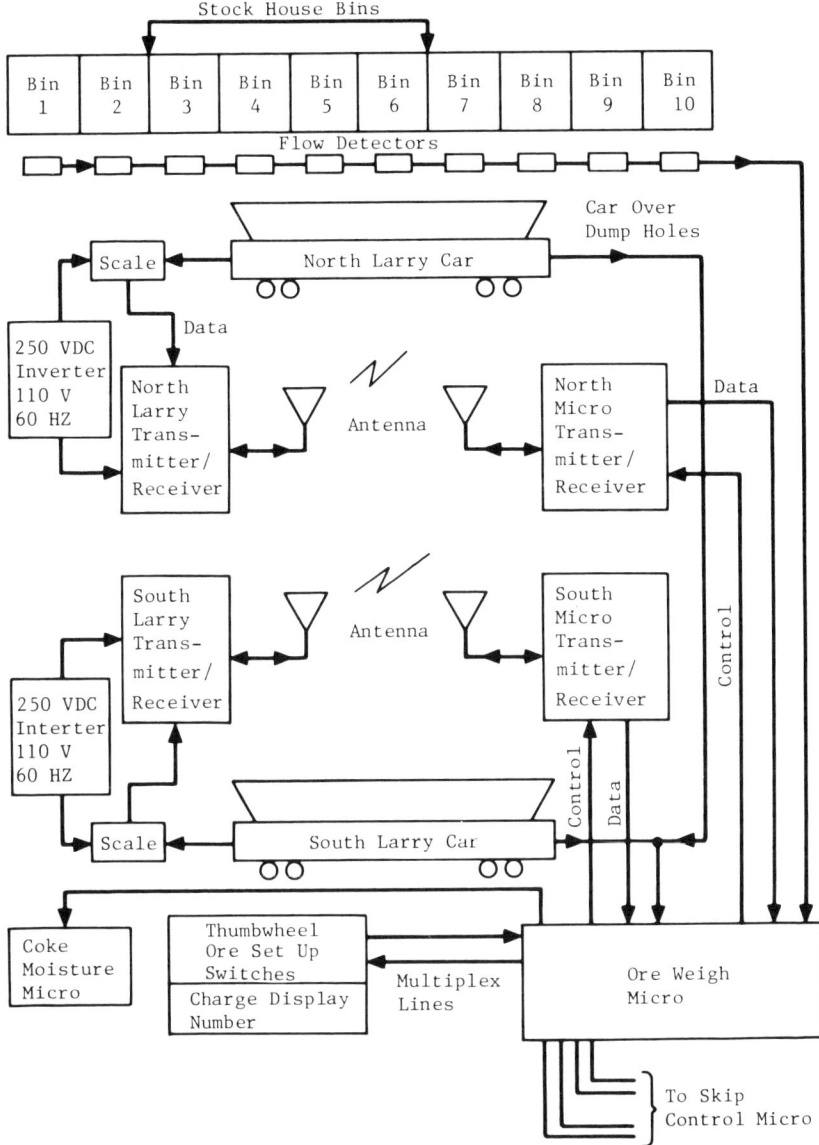

Figure 13. Blast furnace "J" ore monitoring microcomputer system [Farquharson et al. 1978].

Blast Furnace

Blast furnace computer controls are now limited to three main subsystems:

1. burden composition and distribution control;
2. general operation monitoring and data logging; and
3. hot stove switching.

There is no closed-loop control of the blast furnace process anywhere in the world.

Figure 14 shows the relationship between blast furnace instrumentation and process control configuration. To provide the furnace with an optimal burden, the burden composition and distribution at the top of the furnace are controlled automatically. The computer keeps track of raw material characteristics, including the water content of the coke, which can be measured on-line. This computer also establishes daily mass balances.

Efforts are under way to monitor blast furnace operation. Since the core temperature and environment preclude direct monitoring of the fast reaction zone, sensors are limited to peripheral locations, including:

- stock line elevation [Wilson 1978];
- burden surface temperature [Iizuka et al. 1979; Wilson 1978];
- temperature and composition distribution of the gases across the top section of the furnace;
- burden permeability;
- shell movements;
- cooling water temperature [Hill 1978];
- lining wear—monitored in one Japanese furnace [Shiraiwa et al. 1978];
- furnace conditions along its vertical axis (investigated by British Steel Corporation (BSC) research personnel [Craig and O'Hanlon 1978]).

Stoves are designed to maintain a constant hot blast temperature. The stove changing system is operated automatically with provisions for semiautomatic or individual operation of each valve [Malone 1978]. In Inland's No. 7 furnace, the necessary interlocks are part of the computer software [Wilson 1978]. The blast volume, temperature and humidity are automatically controlled.

Two computer systems will be used on Inland Steel's No. 7 furnace. One will consist of control computers and the second will be used for data logging and background calculations [Wilson 1978]. Another, similar, example of a state-of-the-art blast furnace computer system with

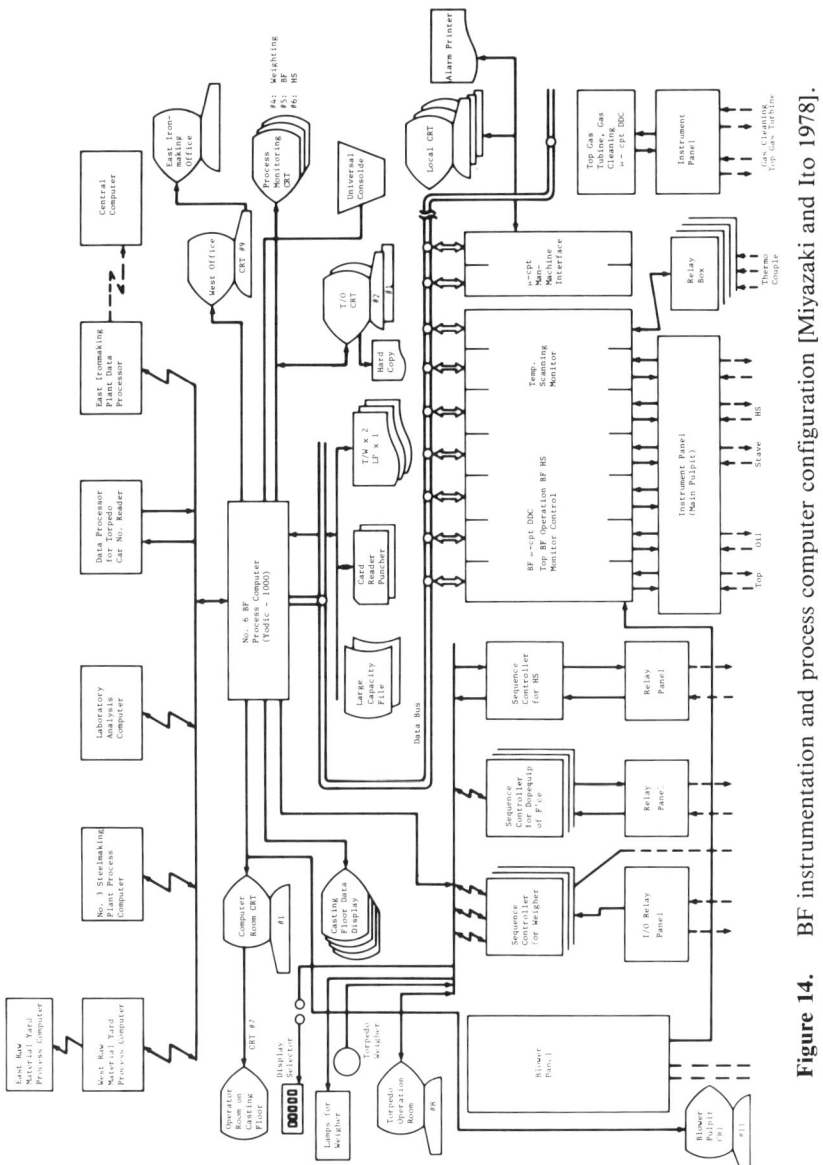

Figure 14.   BF instrumentation and process computer configuration [Miyazaki and Ito 1978].

the same capabilities is operational at Kawasaki Steel Corporation's No. 6 blast furnace at Chiba [Miyazaki and Ito 1978].

Computer systems of the type described here have achieved energy savings of up to 10% of the total fuel (coke, blast furnace gas and injected hydrocarbons) when compared to manually operated furnaces.

A concept of blast furnace control based on a hierarchical system of computers and microprocessors has been described. Although the system has not been implemented fully at any location, it is based on operating systems used in the industrial and power distribution fields. Blast furnace control by computer is made difficult by the lack of an understanding of the processes occurring within the furnace.

The control system is organized into six control areas, each under the supervision of an independent computer and having minimum interconnections with adjacent areas. When malfunctions occur the effect on the other areas is held to a minimum. The control areas are:

1. materials supply: automatic filling of day bins, selection of transportation routes in case of failure and positioning of discharge devices;
2. charging: burden calculation, recipe storage, filling of weighing hoppers, correction of present weights, correction of moisture on coke, tracking of materials, control of furnace top, printing of charge log, and fault annunciation and logging;
3. furnace control: collection of process data, logging and alarms, hot blast temperature control, supplementary fuel injection control, and top gas pressure control;
4. hot stove control: sequence control, process control, stove protection, engineering and production logging, fault annunciation and logging;
5. water cooling: pump supervision, control and logging of pressure, temperature and flow, evaporative cooling, and chemical analysis of water; and
6. blast: control of electric motor or steam turbine.

Since a blast furnace must operate continuously, the ability of the computerized control system to operate in several modes to cope with malfunctions and changing operating requirements is important. A fully automatic mode can be implemented with the process controllers all operational, or with a higher-level computer taking over the functions of a failed processor. A semiautomatic mode is used if the higher-level computer breaks down, or when adjustments must be made to the blast furnace process. Manual control is used to cope with a serious control system breakdown, during startup or during major equipment failure. A local control mode can be used during maintenance and repair operations.

Two different system interactions are used for the basic control system, along with an instrumentation, calculation and reporting system that has provision for communicating to a higher-level computer.

Steelmaking

Raw liquid steel is made either by a BOF, electric arc furnace (EAF) or open hearth furnace (OH). The major subsystems found in BOF control include:

- charge weight control;
- blowing control;
- alloy addition control; and
- sublance control, where applicable.

An example of the state-of-the-art integrated computer control of BOF steelmaking is to be found at the No. 3 BOF shop of the Fukuyama works of NKK, the world's largest steelmaking complex, with an annual steel production capacity of 16 million tons. Figure 15 indicates the main features of the computer control system.

A dynamic control system based on measurements made on the oxygen lance during the treatment process [Kern et al. 1980] has been implemented by Bethlehem Steel on a 270-ton BOF in its Bethlehem, Pennsylvania, plant. The operation started in November 1977. The oxygen probe contains sensors that measure both carbon content and temperature for relay to a process control microcomputer. A sample for spectrographic analysis is also obtained. The process control unit adjusts the oxygen and coolant requirements needed to meet the steel specifications without corrections, reblowing or cooling.

Two computers are used for the process control system. The process computer has been used since startup of the shop in 1968 and has 32K of core capacity and 256K of drum storage. A microcomputer controls the sequencing of the sensor lance.

The process control computer performs the following calculations and functions:

- calculates the charge weight;
- controls the flux weight;
- controls the oxygen blowing rate and volume, and the lance height in the melt;
- initiates the sensor lance test;
- interprets carbon and temperature profiles and displays the results;

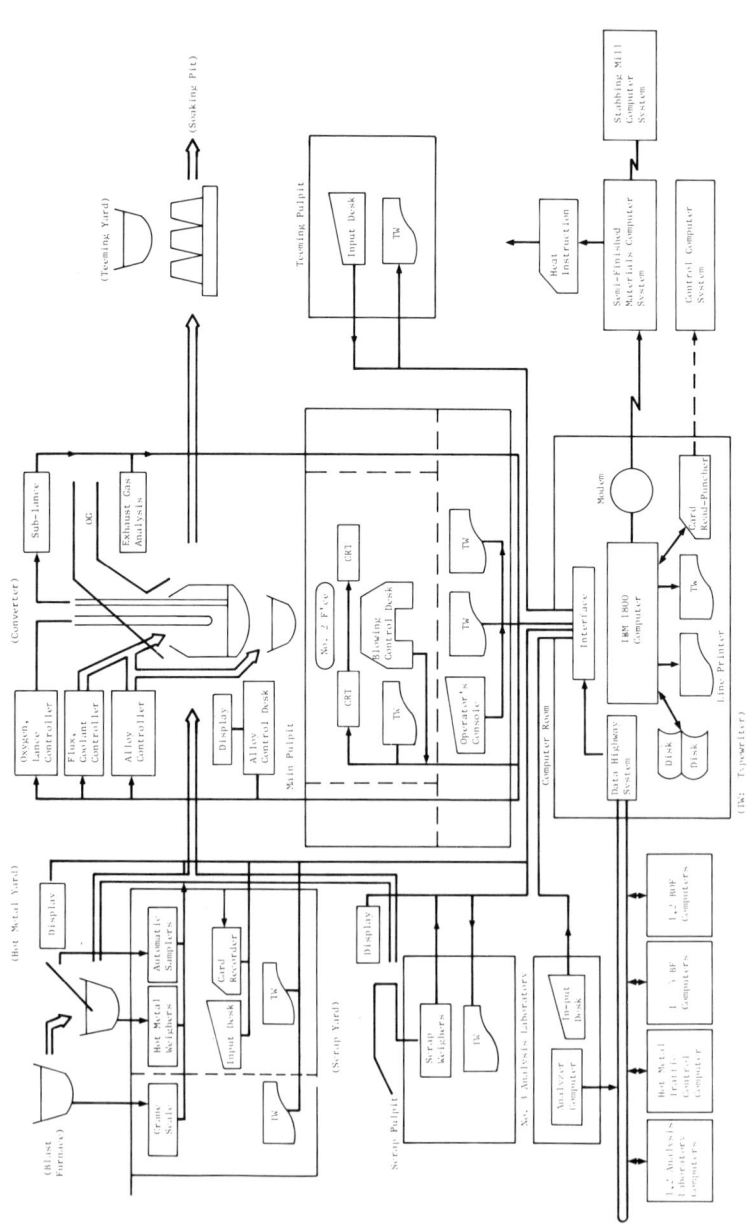

Figure 15.   Computer control system of Fukuyama's No. 3 BOF [Tanaka et al. 1977].

- calculates an adjustment to the oxygen and coolant, and displays the results;
- changes the oxygen and coolant set points if required; and
- initiates the end-of-blow sensor test to confirm the carbon content and temperature.

The microcomputer controls the sensor during the test, and:

- transverses and lowers the sensor;
- sequences measurements and sample collection; and
- raises and transverses the sensor and initiates sample retrieval.

The microcomputer is linked to the process computer by a data link. Spectrographic measurement determines the concentration of carbon, manganese, phosphorus, sulfur and silicon.

Three major subsystems placed under digital computers for the electric arc furnace process are:

1. raw materials mix and weight control;
2. power demand and sequencing control; and
3. steel analysis and alloying addition control.

Other uses of computers in EAF steelmaking ships may include oxygen flow control, raw materials and finished product inventory, in-process product trucking and general data logging. Some time-sharing companies also offer program packages that indicate whether heat may be economically reassigned to a different order, depending on the actual chemistry reached in the bath.

Steel Casting

Currently, computer control, when applied to continuous casting, is supervisory; the computer establishes setpoints for the more traditional analog controllers. Long [1978] reports that, of 25 strands in the United States, 8 are computer-controlled. Current state-of-the-art in computer control involves measurement and control of the mold level which, in turn, establishes strand speed and control of the spray water.

Reheating

A retrofit system described as an interim solution but including many state-of-the-art features is described by Macedo et al. [1977] of BSC for a five-zone slab reheating furnace at the Lackenby works. This is a pusher-type furnace with a rating of 241 ton/hr for 200-m-thick slabs. The control philosophy is based on simply maintaining a consistent slab surface temperature. Features of the computer control system installed in 1975 are illustrated in Figure 16.

Rolling Mills

The automation of steel rolling mills is a good example of the application of process control computers to a complex dynamic operation (Figures 17 and 18). The rolling mill is a flexible production tool that is used to process a variety of different output products from slabs and billets that differ in size, composition and temperature. The stands of a rolling mill must operate at different speeds, but be synchronized to each other and to the roller gap spacing. The thickness of the steel that is produced will vary for a number of reasons. The steel thickness must be measured and the information used to alter the roll displacement while the metal is being processed. The operations of a rolling mill are sequential, and delays in any section hold up operations ahead of and behind the slow process. Since the capital investment in the mill is very high, it is important to maintain high productivity.

Customer requirements for the rolled product are constantly becoming more rigorous. Consequently, it has become increasingly important to have an automated system that can maintain high product quality. During the 1960s and early 1970s great progress was made in applying process control computers to rolling mills, as a means of setting up the mill for new rolling conditions, and adjusting parameters dynamically while the mill is in operation. Additionally, the computer quickly locates malfunctioning equipment and reduces the time required for repair. With standardized rolling conditions, product quality is improved. Another advantage of automation has been that the operator has more time to observe the process and other problems. Adjustments of the process and malfunction detection are performed more quickly.

In automating a five-stand cold rolling mill, the South African Iron and Steel Industrial Corporation was initially motivated by a desire for labor savings, but experience with the system has shown that consistent product quality is the most desirable result. Standardized rolling conditions improve the quality and consistency of the project. Another desirable benefit of automation is the rapid detection of equipment malfunctions by the operator. This increases the amount of time that the mill is in production. The loss in revenue that occurs when a rolling mill is

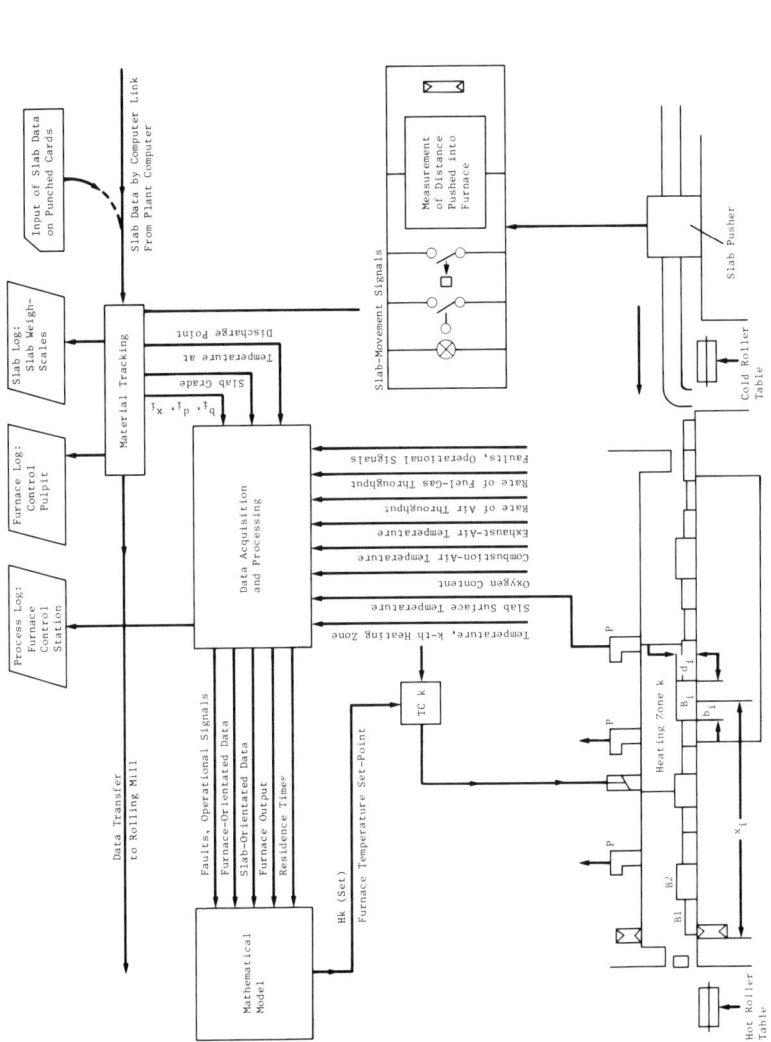

Figure 16.   Automatic furnace control using a process computer. P = pyrometer; B = slab; TCk = temperature control of zone k; b_i, d_i = dimensions of slab; x_i = position of slab [Buchmann 1978].

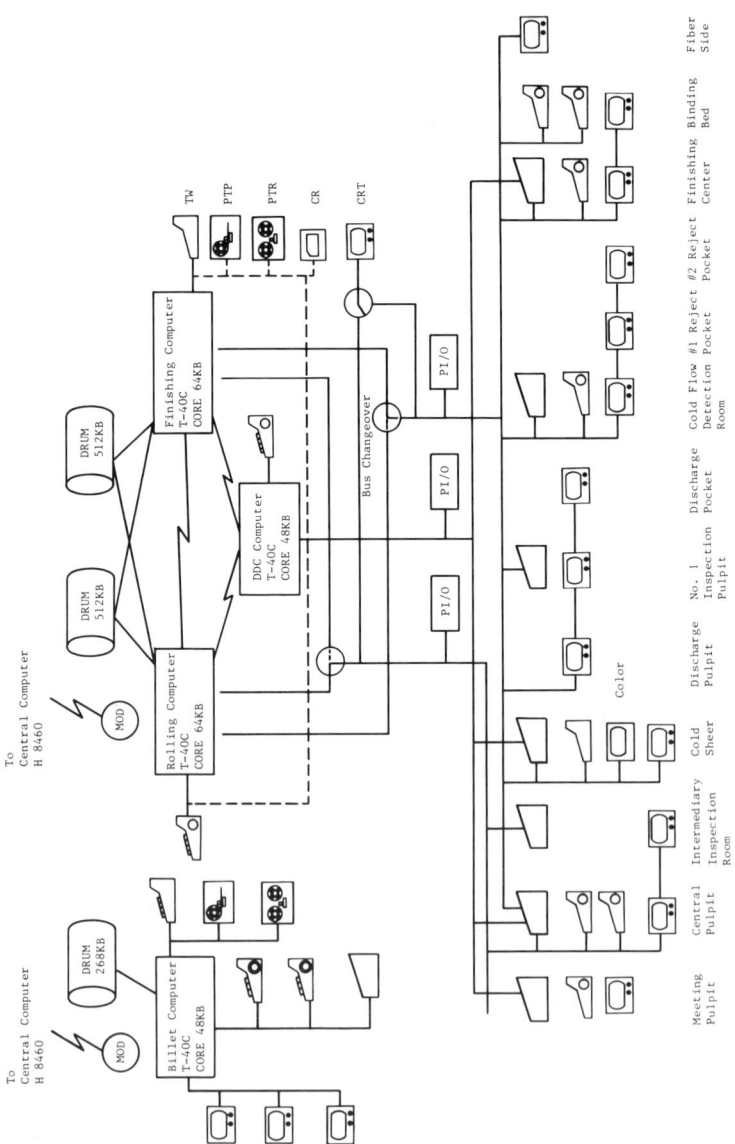

Figure 17.   Block diagram of the computer control system for a bar mill [Kamii 1976].

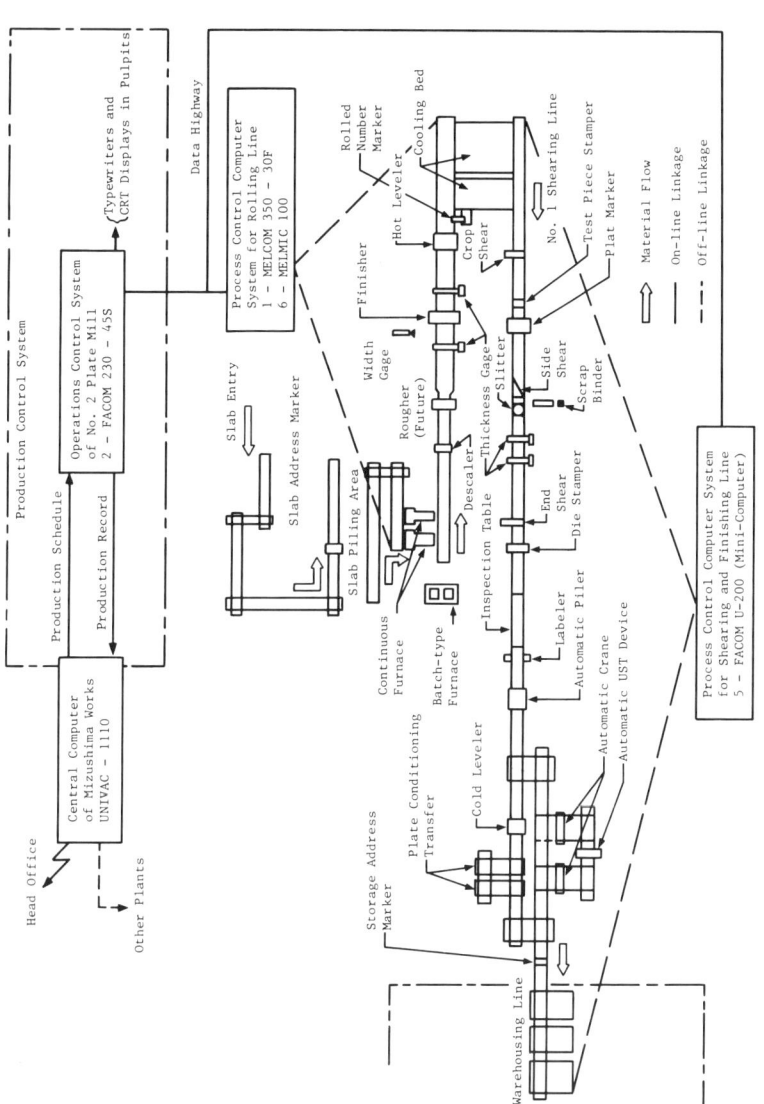

Figure 18.   Total automation—No. 2 plate mill [Lida et al. 1978].

down for repairs is $10,000/hr. This penalty for nonproductive time also caused the conversion of the mill to computer control to be stretched out over a long time because the installation of the new equipment had to be scheduled during the idle periods occurring during normal mill operation.

In normal operation, reference values of rolling speed, sheet tension, roll position and force are calculated for each coil and relayed to the lower-level control system. The stands are adjusted to the starting condition and a dynamic control system is used to adjust the parameters around the reference condition. Tension between stands is controlled, and all stand roll forces are independently controlled by fast-response control loops during rolling. The functions of the computers in the system are:

1. optimizing computer: distribution of total thickness reduction over five stands, calculation of all reference values from mathematical models, statistical evaluation of measured values, data logging, and roll bending;
2. minicomputer 1: acquisition of measured values during rolling, automatic gauge control, roll position control, and mill slowdown at end of a coil;
3. minicomputer 2: sequencing mechanical movements and roll changing cars, and generation of the master ramp signal for all main drive speed regulators; and
4. minicomputer 3: sequencing all entry and delivery mechanical movements of slabs and finished coils, and coil diameter and width measurements.

The production data that are automatically logged are:

1. production report on each coil: initial and final weight, rolling time, total length, and classification of length into thickness bands;
2. production report for each shift: total weight and number, tons per hour, percentage coils needing rework, and total delay time;
3. engineering report: precalculated values, measured values, confidence intervals for all parameters, and production parameters for a single coil;
4. fault log: coil history for 12 sec before a break; and
5. alarm log: drive bearing temperature, strip tension, and hydraulic pressure.

Rolled products are manufactured in a wide variety of sizes and shapes, yet all rolling mills share common automation problems. Each control area may be more or less emphasized, depending on local conditions. The state-of-the-art of plate mill computerization is to be found in Japan

[Takeuchi et al. 1977], while several examples of modern, automated hot strip mills may well be found in the United States, e.g., at Inland Steel or Armco, Inc., in Middletown, Ohio [Long 1978]. Less computer emphasis has been reported on bar mills or structural mills, perhaps due to smaller volumes of product or a wider acceptable range of dimension specifications for acceptable product, e.g., reinforcing bars.

Subsystems where computers have been applied in rolling mills include:

- mill scheduling;
- identification and materials tracking;
- automatic gauge control;
- motor drive and interstand tension control;
- automatic cutting and shearing; and
- aspects of automatic quality control.

Heat Treating and Finishing

The state-of-the-art computer control of annealing furnaces is illustrated by the furnace firing dynamic DDC developed and implemented by Armco, Inc., at its Middletown, Ohio, works. The approach adopted is the determination of firing time as the box is being fired. An on-line soak time model monitors bottom coil temperature and predicts the temperature of the coldest spot in the coil, which in turn is used in determining the completion of the soak time. The model is based on theoretical heat transfer considerations, but can be fine-tuned to reflect specific furnace and coil conditions based on historical data.

Fully Computerized Seamless Tube Mill

A specialty tubing mill installed at Ambridge, Pennsylvania [Laird et al. 1980], is a modern, flexible facility designed to satisfy the changing demands of the specialty marketplace. This combination mill is a mandrill mill producing small-diameter tubing and a transval mill producing bearing and mechanical tubing. A rotary furnace has a capacity of 75 ton/hr; a reheat walking beam furnace has the same capacity.

Two control computers are used to sequence the mill in a manner unique in the seamless tubing industry. The computers supply all drive references, perform all position regulator functions, and support 14 CRT at the operator's stations. Each computer has a backup central processing unit (CPU) to minimize delay during malfunctions. One computer system controls all equipment from the rotary furnace to the reheat furnace. The other computer controls from the reheat furnace to the outlet. Except for tracking information passed between computers, each system is independent.

The CRT are used to display 57 formatted items of information used in mill setup, tracking and monitoring. Product size and specific operating conditions are entered. From these, main drive speeds, crewdown settings, table adjustments, and entry and exit speeds are calculated. Calibration displays are used to recalibrate 62 regulators for roll wear, position verification and checks after maintenance work. Mathematical models in the computer calculate setup parameters.

The computer system tracks billets from the cold billet loading dock to the end of the mill. The number of units processed and those remaining at any stage of the process are displayed to operators on the mill floor. When maintenance is required, the entire collection of 4500 schematic diagrams can be displayed on any CRT in the mill to aid in repairing the equipment from the most convenient location. The computer system consists of four Westinghouse 2500 process control computers with 89K of 16-bit words and a 500,000-word fixed-head disk memory. The primary software components consist of the graphic sequencer, position regulator, CRT console monitor and mill preset programs. The graphic sequencer senses inputs from the mill, performs logic analyses, and issues commands to the mill. The position regulator adjusts the mill drives. The console monitor provides interface data to the operators. The mill preset programs allow mill setup to commonly used rolling schedules using main drive motor speeds, positioning regulated drives and moving data from the preset area to the current area.

Two of the operating personnel spent two years working at Westinghouse while the computer control system was being developed. If more people had been used, the time required to start up this complicated plant would have been reduced. Additional personnel who understood the system would have reduced the startup costs.

Induction Heating

An electrical induction heating system [Craig 1979] to heat steel slabs prior to entry into a rolling mill was built in 1969 by McLouth Steel in Detroit, Michigan. The coils are large enough to handle a slab of steel 12 × 60 in. (0.35 × 1.52 m) × 26 ft (7.9 m) in length, and weighing 30 tons (27.3 metric tons). The furnace has a capacity of 600 ton/hr (545 metric ton/hr). The furnace consumes 210 MW of electrical power and must be operated from the local utility, with power purchased during periods of low demand.

The induction furnace was selected over a fossil fuel design because of initial cost, high yield, minimum maintenance costs, suitability to computer operation and low operating cost at the time of installation. The system consists of a 120-kV incoming line, main power transformer, power factor connecting capacitors and the inductive coils. The bus bars are water-cooled and supply current to four sections of coils that completely enclose the steel slab. The power can be varied in each coil in 4% increments, to heat the slabs uniformly and hold the slabs at temperature during delays.

Slab-handling cranes and table roll motors are controlled by computer. The slab's temperature is raised in successive positions in the furnace and kept at the proper temperature until the reversible rolling mill is ready to receive the slab. The computer regulates the current in the coils to the heating requirement and the total electrical load of the entire plant.

The facility heats an average of 2225–2350 tons (1109–1166 metric tons) per 8-hr working shift, using about 340 kWh/ton (384 kWh/metric ton). When the facility became operational in 1969 the heating cost was $2.92/ton ($2.65/metric ton); by 1977 the cost had increased to $8.65/ton ($7.85/metric ton). In the early days of operation many mechanical and electrical problems were encountered, but failures have decreased and the unit is now very reliable. Coil sections are replaced after 5–9 years of service.

The energy and environmental benefits of induction heating are:

- no air pollution;
- no water pollution;
- low noise level;
- energy ultimately supplied from coal;
- facility shut off when not in use; and
- facility in use during off-peak hours when utility has excess generating capacity.

Steel Angle Bar Punching Machine

The application of a microcomputer to the control of a steel angle punching machine [Horbal and Derrick 1978] represents a specialized example of the control capability of these devices. The machine to be controlled consists of four hydraulic punches that produce holes in steel angle stock for a variety of applications. Actuators to move the stock through the machine are also controlled by the system. A modular design approach uses the microprocessor to supply control signals to a set of solid state hydraulic control relays that actuate the punches. Hydraulic servos position the steel stock in the proper location for actuation of the punches. Other functions interchange punch dies, rotate the punches and place an identifying mark on the product, for a total of 11 independent operations.

The operation is controlled from a CRT-equipped panel used by the operator for interaction with the computer. Control programs and algorithms are written in the MACRO-11 assembly language, which is fast enough for real-time control. The memory consists of 8K words of 16-bit programmable read-only memory (PROM) assisted by semiconductor random-access memory (RAM). The PROM resident software consists of a:

- monitor kernel;
- editor and compiler;
- main control program;
- datalink program; and
- diagnostic package.

The advantages of the computerized system are greater use of machine time, reduction of bad parts by computer checking of programs and better communication with the operator during setup.

Maintenance of Microcomputers in the Steel Industry

Because of the exposure to temperature extremes, dust, shock, vibration, electromagnetic radiation pickup and other detrimental conditions, a steel mill is a very demanding environment in which to operate microcomputers. The industry has devised its own designs for successfully mounting equipment in this environment using special air-conditioned rooms required for environmental controls as well as special mounting procedures for use under harsh conditions. Trained crews must maintain the equipment to minimize computer failures.

Despite the harsh operating conditions, microcomputers can serve at least two useful functions in steel mills: automatic skip control and ore monitoring for a blast furnace. Cichelli [1979] discussed the special problems related to such a computer installation, noting that a special interface relay panel was built to avoid long cable runs of low-level signals.

Special attention was paid to isolating lines from electromagnetic noise and isolating signal lines with good wiring and isolation practice. Hardware alarm lights are extensively used because 100% availability is required and it is necessary to correct any failure as rapidly as possible. Self-check features are programmed into the microcomputer software to assist in locating the failure. The difficulty is often in external equipment, rather than in the microcomputer, and is often a physical disturbance rather than an electrical problem. Check sum numbers are used to quickly establish the validity of the microcomputer program. At the test steel mill discussed by Cichelli [1979] equipment and personnel were trained with their own program.

Steel Plant Modernization

The modernization problems of the steel industry have been carefully considered in plans by the United States Steel Corporation to build a completely new steel mill at Conneant, Ohio, on Lake Erie near the Pennsylvania border. This plant is planned to produce 3.4 million metric tons of steel per year, to employ 8000 people and cost $3 billion. It will be the first totally new steel mill constructed in the United States since 1961. The cost of this one plant will amount to more than the entire steel industry in the United States invested in capital equipment in 1975 and 1976, and will add 10% to the productive capacity of U.S. Steel.

Another example of the large investment required to build a totally new steel plant is the new plant at Nanticoke, Ontario, being erected by the Steel Company of Canada. Requiring six years from the planning stage to initial production, the plant represents an investment of $1 billion.

The advantage of constructing a completely new mill is that the components can be of maximum size for maximum economy of scale, and will be properly matched in size to preceding and following operations. The 8-m-high coke ovens and 14.4-m-diameter blast furnace will be among the largest in the industry. The predicted energy consumption will be 30% below the average current U.S. practice; productivity is expected to be increased by 40%. The installation of new equipment in existing mills often does not result in a good match to the productive capacity of the older equipment. In addition, factors such as lack of space force compromises with the best design.

Since coke ovens already operate at 90% efficiency, the biggest potential for improvement and energy savings is obtained by using dry quenching, which produces 0.5 kg steam/kg coke. The blast furnace, which consumes almost half of the energy used in steel making, is 80% efficient and can only be improved by 2–3%.

The biggest improvement in the Ohio plant will come from adopting continuous casting. Presently, 10% of steel in the United States is continuously cast. Rolling mills waste the energy used for reheating if they are used for stop-and-start operations. Continuous operation makes the scheduling of numerous small orders difficult, however, and leaves no time for inspecting and conditioning slabs.

The Ontario plant will consist initially of a loading dock for ships, raw materials storage yard, one blast furnace, two BOF and a slab casting facility. A coke oven battery and rolling mill will be added later. At first, coke and steel rolling will be provided by another plant located nearby. The plant has been designed to environmental standards that are as rigorous as those in the United States. A rolling mill that can produce wide, heavy coils would have the capacity to roll 4–5 million ton/year and would cost $500 million. The coke ovens will be the largest in North America, standing 6.7 m high. The plant will have an initial capacity of 828,938 ton/yr (752,000 metric ton/yr) and will be rapidly expanded in several additions to 1,286,384 ton/yr (1,167,000 metric ton/yr) [*Iron Age* 1980].

The initial production is below that of an optimum-sized plant and will not be cost-competitive until a production of 1.8 million ton/yr is achieved. Ultimately, expansion to 6 million ton/yr, and then 12 million ton/yr will occur.

It is significant to observe that the construction of the Ohio plant has been delayed because of unfavorable tax laws in the United States, while the construction of the Canadian plant has begun. The two plants would be located on opposite sides of Lake Erie, yet, because of the difference in tax depreciation, the Canadian plant is projected to have a return on investment one-third larger than the American plant. During summer 1980 the Canadian plant was under construction, but the American plant was not being built because it could not be justified economically.

Computerized Test Method

The use of a microcomputer to process acoustic signals was reported [*Iron and Steel Engineer* 1979] as a nondestructive testing method. Pipes and tanks that are to be pressure-tested are filled with water and slowly pressurized. Transducers attached to the steel structure under test record the acoustic emission of the metal in response to the accumulating stress in the metal. The acoustic signal is processed by a minicomputer to detect the signature of an expanding crack and record the test data for a permanent record. The computer can categorize the severity of the crack as it grows; three transducers can determine the position of the defect by comparing the exact time that the sound reaches the transducers. The test data is analyzed in the field and is performed in several hours so the object under test is out of service for a minimal time.

Hierarchical Computer Control Systems

In 1973 the Purdue University Laboratory for Applied Industrial Control (West Lafayette, Indiana) began a four-year project to develop the specifications for an overall hierarchy computer control system for a steel mill [Williams 1980a]. Because of the size and complexity of the project, the study was extended and is continuing with the active support of eight steel companies. Teams with industrial and academic backgrounds were set up to study the tasks to be performed at each level.

In addition to the specification for the computer control system, the following results were also produced:

- a steady-state blast furnace model;
- an ingot heating and cooling model permitting on-line optimization;
- soaking pit and slabbing mill models;
- an overall steel mill simulation model suitable for scheduling activities;
- a new technique for the adaptive control of rolling mill setup;
- a reliability study simulation;
- an in-plant rail transportation system simulation;
- a study of communications techniques between computers;
- a study of on-line defect detection of steel products; and
- simulation techniques for manpower requirements.

Originally, a four-level hierarchy was proposed consisting of:

1. Specialized dedicated digital controllers will interact with discrete process steps and be controlled by the operators at level 2.
2. A direct digital control level will display data to the operators and receive commands. In addition to having direct links to the process and supervisory control over the level 1 controller, the level 2 units exchange information with other level 2 units and the level 3 controller.
3. The supervisory control level displays data to supervisors, receives their commands and communicates with units at levels 1 and 2.
4. The management information level is the highest level in the hierarchy. It interacts only with the supervisory level 3 and relays the commands of management to the lower levels.

Only one level 4 system would control an entire plant. A single level 3 system might be used to control all hot rolling processes. Separate level 2 systems might be used for reheat furnaces, roughing and finishing mills.

It is impossible to analyze accurately the economic potential of adopting a hierarchical computer system in the steel industry. Experience in Europe and Japan (Figure 19) with less sophisticated systems indicates that the benefits are large, but the necessary data are not available, and the economic evaluation would itself require a large-scale study. An important parameter in the economic evaluation, the personnel requirement, has not been determined.

Ferrosilicon Alloys Manufacturing

An entire ferrosilicon alloys manufacturing operation (at the Ashtabula, Ohio, plant of the Metals Division of Union Carbide Corporation) has been converted to operate under closed-loop process computer control.

In this installation, a single process control minicomputer is used to completely control the operation of two 50- to 55-MW ferrosilicon furnaces and their associated auxiliary subsystems. The closed-loop control extends from the metallurgical calculations of the mix batch composition, and mix batch weighing and delivery, to the final analysis of the alloy as it is tapped from the furnace. The total furnace plant operation is divided into the following separate process modules:

- raw material batching and delivery;
- metallurgical calculations and corrections;
- electrode consumption and slipping;
- electrical power input and regulation;
- tapping interval;
- offgas and fume collection;
- alarm condition monitoring and correction; and
- data monitoring and logging.

The raw materials batching and delivery systems consist of 34 bulk storage bins, apron conveyors, vibrator feeders, hydraulic load-cell weight modules, conveyor belts and batch delivery cars. Radioisotope level sensors monitor the level of mix in these bins and signal the computer when more mix is required.

Once per minute, the computer scans the level sensors on each furnace. When mix is required for one or more furnaces, the computer engages simultaneously in three processes: (1) determining which bin should be given priority based on the rate of use from each bin; (2) determining the

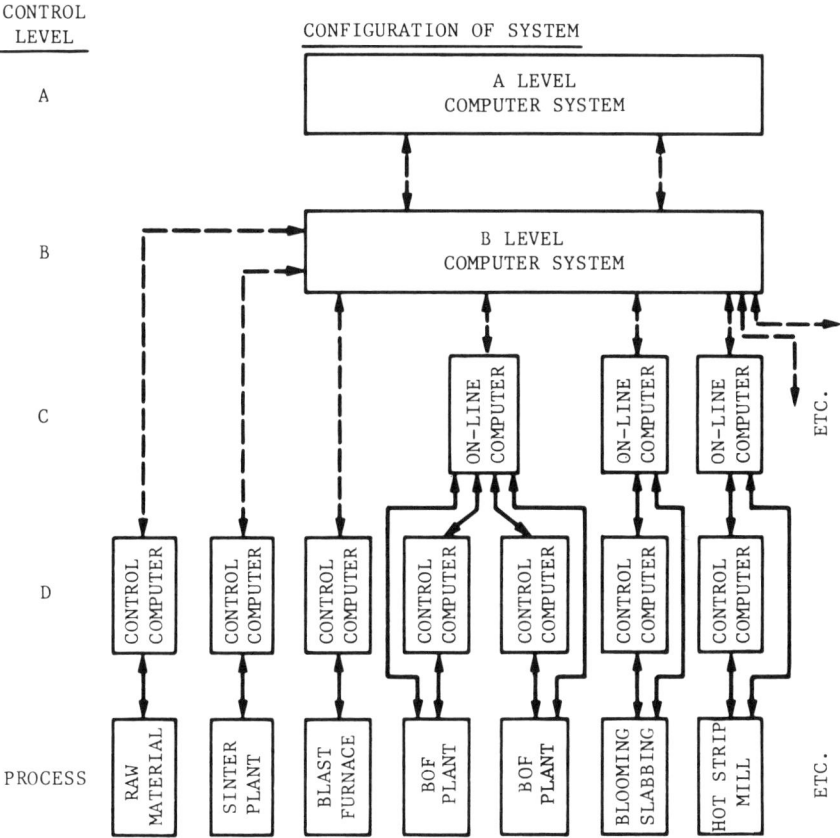

CONTROL LEVEL

CONFIGURATION OF SYSTEM

A — A LEVEL COMPUTER SYSTEM

B — B LEVEL COMPUTER SYSTEM

C — ON-LINE COMPUTER ... ETC.

D — CONTROL COMPUTER

PROCESS — RAW MATERIAL, SINTER PLANT, BLAST FURNACE, BOF PLANT, BOF PLANT, BLOOMING SLABBING, HOT STRIP MILL ... ETC.

Figure 19. All on-line production and process control system at Kimitsu Works, Nippon Steel Corporation, Japan. *Control level A:* span—whole works; cycle time—10 days/daily; major functions—order processing, order modifications, material requests, production scheduling, production completion, shipment, instructions and reporting. *Control level B:* span—plant; cycle time—daily by shift; major functions—collection of mill data, production scheduling, collection of test data, production allocation. *Control level C:* span—production area; cycle time—on-line, real-time; major functions— production instructions, production reports, product progress. *Control level D:* span—process equipment; cycle time—on-line, real-time; major functions— control of process, operation instructions, quality control data, engineering data.

raw material bin and quantity of each raw material; and (3) transmitting the batch request to a programmable controller. The programmable controller verifies the batch request and weighs the raw materials simultaneously into the weight modules. The controller transmits the actual weight of each material back to the computer, which logs the data and performs a weight comparison to provide correction for weighing error. Should a raw material bin run empty, the computer will direct the controller to weigh from another bin containing the same raw material, or from a bin that has a permissible substitute. After the weighing is complete, the raw materials are delivered to a common hopper, where the total weight is compared to the sum of the individual weights. Once the check weight is performed, the batch is delivered to the furnace.

Metallurgical control is primarily based on raw material analysis, product analysis and electrical characteristics of the operating furnace. Each tap from the furnace is sampled. The sample is sent to the plant laboratory for X-ray analysis. The results are then relayed directly back to the furnace computer from the X-ray spectrometer. Taking trends into account, the furnace computer compares actual to desired analyses and adjusts the batch composition of the raw materials to the furnace. Raw materials are also adjusted if the electrical characteristics indicate an impending poor furnace condition or unbalanced operation.

Using its ability to predict what will happen when the voltage is changed on a transformer, the computer regulates the power input without exceeding the electrical limits of the current-carrying components or the electrodes. Metallurgical adjustments are imposed if maximum load or balance cannot be obtained through voltage regulation. Using optical pyrometers, the computer senses whether the furnace is being tapped. A warning light system, operated by the computer, signals tapping personnel when a tap is impending and when to open the taphole. The computer logs the tapping when it sees the taphole opened and closed. Automatic load reduction is imposed if the taphole is not opened within limits, and eventual shutdown will occur if the limits are radically exceeded.

All critical components of the furnaces are monitored by analog or digital inputs. Any analog input can be flagged for alarm action on high or low limits or a rate of change, and any digital input can be set to trigger alarm action in the open or closed position. An alarm condition or failure of any critical component results in immediate power reduction to the furnace or shutdown. The alarm restriction is not removed by the computer until the alarm condition is cleared.

The degree of improvement that can be attained through the application of computer control naturally depends on the quality and uniformity of operation from which one starts. The comparative results (Table XIII) for two furnaces reflect this difference. Before addition of computer control, Furnace No. 20 had experienced several years of excellent performance on 50% ferrosilicon operation. On the other hand, following its initial startup, operation of Furnace No. 23 had suffered

Table XIII.   Operating Results of Union Carbide Arc
Furnace Computer Control [Ray and Wilbern 1979]

Key Operating Parameter	Improvement Under Computer Control	
	No. 20 Furnace	No. 23 Furnace
Operating Time	3.3%	3.3%
Average Operating Load	6.0%	8.3%
Kilowatt-hours per Pound of Alloy	1.4%	3.0%
Pounds of Electrode per Pounds of Alloy	2.0%	2.3%
In-grade Production	1.9%	0.9%
Labor Hours/per Ton of Alloy	13.3%	13.3%
Tons of Alloy	10.6%	14.1%
Cost per Pound of Alloy	3.6%	5.2%

through an extended period of mechanical equipment problems, electrode failures and operator learning curve requirements.

Through providing a more stable furnace operation, the computer has reduced production costs and increased productivity. A constant electrode penetration in the furnace, along with a highly successful computer recovery technique following furnace outages, has resulted in a greater operating load, more operating time, less power use per pound of alloy, and less use of electrode per pound of alloy. More frequent and consistent metallurgical analysis of the furnace and more accurate raw material weighing have resulted in a higher percentage of salable product. Maximizing power input and balance has resulted in more tons of alloy from the furnace. A major unquantified benefit has been the improved safety for the operators.

The total cost of the computer amounts to 20 man-years of effort or about $1.2 million. About $450,000 is in program development, which need not be repeated. The installation can operate about 99.5% of the time under computer control.

Economic Justification of an Automated Machining System

Cincinnati Milacron, Inc., the largest U.S. manufacturer of machine tools, has performed a comparison between the economic feasibility of an automated machining system and a traditional operator controlled set of tools [Holmes 1979a,b]. The automated system consists of two automated horizontal lathes serviced by a robot for loading and unloading, and a conveyor for feeding stock and receiving finished workpieces. The manual system consisted of three ordinary lathes with their operators.

Productivity was higher with the robot-fed automatic lathe because of more rapid loading and unloading, accentuated by the fact that the human operator slowed down as he became fatigued. A slightly larger capital investment was required for the automatic system. The two-lathe and robot automated production cell cost 7% more than manned lathes. However, operating costs were lower because the automated cell requires only one operator while the manned lathes require three. Detailed cost analysis showed a discounted cashflow return on investment of 65%. A more substantial gain would be obtained if the equipment were operated for more than one shift.

Additional benefits of the automated system not quantified are:

- reduced scrap losses;
- reduced inventory costs;
- better scheduling procedures; and
- quicker production run changeovers.

The cost of an automated machining system is high. The system must be fed a constant stream of raw stock, which is processed at a rapid throughput for 8 hr/day, 5 shifts per week, and preferably 10 shifts per week or more. Few shops have the workload and manufacturing control system to justify this high rate of productivity. Less than 3% of the machine tools now sold by Cincinnati Milacron, Inc., have automatic controls.

Other Applications

The metal fabrication industry, because of its size, diversity and complexity, provides ample opportunities for automation. Automated workpiece holding and handling technologies have begun to have an impact on the way machine tools are designed and used [Arneson 1979,1980; Larsen 1979]. There are 1000 die casting plants in the United States with a combined $4 billion/yr in sales. A manufacturing system unites machines and operators into a single high-speed manufacturing process [Bennett 1979]. In addition, a rail-making facility in Pueblo, Colorado, features computer control in its operation [Cathey 1979].

CHEMICALS

Industry Characteristics

The chemical industry is vast. It is organized by the Census Bureau as shown in Table XIV [DOC 1980a–g]. The chemical industry is the most energy-intensive of the manufacturing industries classified by DOC with two-digit codes. It is unique in using energy materials both as fuel and as feedback or raw material.

Total purchase of fuel and electricity used as sources of energy by the chemical and allied products industry amounted to 3.02 quadrillion Btu equivalents (3.17×10^{15} J) in 1976. The breakdown of energy requirements by energy type for the year 1978 is shown in Table XV. Natural

Table XIV.   Organization of Chemical Industry [DOC 1980]

Sector	1977 Value of Shipment (10^6 \$)	Change from 1972	1977 Employment (1000s)	Change from 1972
Industrial Organic Chemicals				
2861 Gum and Wood Chemicals	391	+18	4.8	+19
2865 Cyclic Crudes and Intermediates	5,637	+175	35.7	+27
2867 Industrial Organic Chemicals, NEC[a]	24,233	+163	112.3	+10
Agricultural Chemicals				
2873 Nitrogenous Fertilizers	2,602	+222	12.1	+29
2874 Phosphatic Fertilizers	2,862	+127	14.4	+3
2875 Fertilizers, Mixing Only	1,867	+133	12.4	+9
2879 Agricultural Chemicals, NEC	2,790	+142	15.0	+23
Industrial Inorganic Chemicals				
2812 Alkalies and Chlorine	1,655	+101	11.8	−11
1813 Industrial Gases	1,235	+82	7.5	−22
2816 Inorganic Pigments	1,260	+58	11.9	−7
2819 Industrial Organic Chemicals, NEC	8,616	+125	78.2	+23
Miscellaneous Chemical Products				
2891 Adhesives and Sealants	1,808	+95	166	+11
2892 Explosives	666	+65	12.3	+34
2893 Printing Ink	945	+86	10.1	+5
2895 Carbon Black	468	+106	2.5	−14
2899 Chemical Preparations, NEC	3,951	+79	35.3	−5

[a]NEC = not elsewhere classified.

Table XV.   Chemical Industry Energy Requirements:
Fuel, Power and Feedstock (1978 industry estimates[a])
[DOC 1980a-g]

Energy Type	(Trillion Btu) Quantity
Natural Gas	
As fuel	1,613
As feedstock	567
Subtotal	2,180
Petroleum	
As fuel	452
As feedstock	
Heavy liquids	1,154
Natural gas liquids	901
Subtotal	2,507
Coal	318
Electricity[b]	1,736
Total	6,741

[a]"Petrochemical Industry Profile—1978," Arthur D. Little, Inc., Cambridge, Massachusetts, 1979.
[b]Purchased power; excludes internally generated electricity.

gas and electric power accounted for a large portion of total purchased energy.

When the energy equivalent of petrochemical feedstocks is added to that of purchased fuel and electrical energy, the chemical industry has appreciably greater total energy requirements than any other manufacturing industry. Table XVI summarizes the energy requirements of this industry for both fuel and feedstock, with a breakdown by energy form.

Of the 6.74 quadrillion Btu equivalents (7.10×10^{15} J) of energy required by the chemical industry in 1978, 38.8% (2.62 quadrillion Btu or 2.76×10^{15} J) was consumed as feedstock for the manufacture of petrochemicals. Natural gas used as feedstock represented 26% of all gas used by the chemical industry; while 82% of the petroleum and gas liquids were used as feedstock.

The efficiency of energy use for chemical manufacturing has steadily increased since initiation of the voluntary conservation program in the early 1970s. As of the year ended June 30, 1979, the net energy consumption rate had been reduced by 20.9% per unit of output, as compared with the benchmark year 1972.

Table XVI.   Chemical Industry Energy Use by Type[a] (1978)
[DOC 1980a-g]

Energy Type	Percent of Total
Natural Gas	37.0
Other Gas	10.5
LPG – Propane	0.5
Distillate Fuel Oil	1.9
Residual Fuel Oil	8.0
Coal	8.8
Electric Power (purchased)	26.2
Steam (purchased)	3.5
Other	3.1
Total	99.5

[a]Energy as feedstock excluded.

Overview of Automation in the Chemical Industry

The chemical industry is a vast, diversified sector. Chemical plants can be continuous or batch plants. In a continuous plant, each piece of equipment is designed for a specified product and throughput. In contrast, batch processes are carried out using standard equipment under operating conditions that can be readily adjusted to handle a variety of feeds and final products [Brodman and Smith 1976; Mauderli and Rippin 1980].

Chemical plants are automated to ensure higher performance and productivity as well as for energy savings. Automated process controls have been reported as used for:

- ethylene [Funk 1980; Hammett and Lindsay 1976; Skrokov 1976];
- ammonia [Diagre and Wieman 1977; Gremillion 1979; Weems 1979; Yost et al. 1980];
- refinery [DiBiano 1981; Latour 1976; Ritchey et al. 1976];
- polyvinyl chloride [Kennedy 1975];
- plastic [Greene 1981a,b,c];
- lube oil [Spellman and Quinn 1975]; and
- distillation [DOE 1979; Seeman and Nisenfield 1975].

In many cases automation and process control are achieved with digital computers, but sometimes they cannot be used. In a number of cases pneumatic control systems have proven successful at a lower cost than would have been possible with digital control [Gremillion 1979].

Ethylene Plant

A typical atmospheric crude tower in an ethylene plant (Figure 20) yields 20–40% of its products as vacuum gas oil, which is usually priced below crude oil. In addition, a full 1% of the fuel oil equivalent (FOE) is consumed if the charge furnace sections of the plant are not coordinated. For a plant processing 12.3 million kg/day of crude, this translates into 2.46–4.92 million kg/day of feed converted into vacuum oil, and 123,000 kg of feed used to fire the furnace.

Implementing automatic process control in an ethylene plant results in significant benefits because of the ability of the system to always operate near the optimum parameters. Automatic control of the charge furnace cuts fuel consumption by about 2% in a well sealed and maintained furnace. For the plant discussed above, it is possible to reduce vacuum gas oil production by 1–3%. At the same time, this reduction in yield of low-priced products results in an increased quantity of more valuable products. The net profit is increased by reducing waste and fuel use, and by increasing the yield of the more valuable products.

Ammonia Plant

More than 40 computer systems have been installed in ammonia plants worldwide [Gremillion 1979], attesting to the benefits associated with automation in chemical plants. Modern ammonia plants operate under conditions requiring precise process control because many disturbances can affect their normal operation. The feed gas rate may be changed several times a day to fully use, but not exceed, the daily quota of gas. Changes in the primary reformer operation affect methane leakage into the synthesis loop. These changes in methane leakage will require changes in purge rate to hold the desired loop pressure of level of inert gases. The change in purge rate introduces an additional disturbance in the primary reformer, where the purge is burned as fuel. The extensive heat exchangers incorporated in a modern ammonia plant are able to transmit the disturbances throughout the entire plant. The efficiency of the turbines driving the air, makeup gas, and recycle gas compressors shift due to weather and day-to-night changes. Even the best human operators sometimes find it difficult to know when and how much correction to apply to compensate for a change in feed or purge rate.

In 1976 the Cominco American, Inc., ammonia plant in Borger, Texas, was switched from manual to computer control. Automatic control was implemented in the following areas of operation:

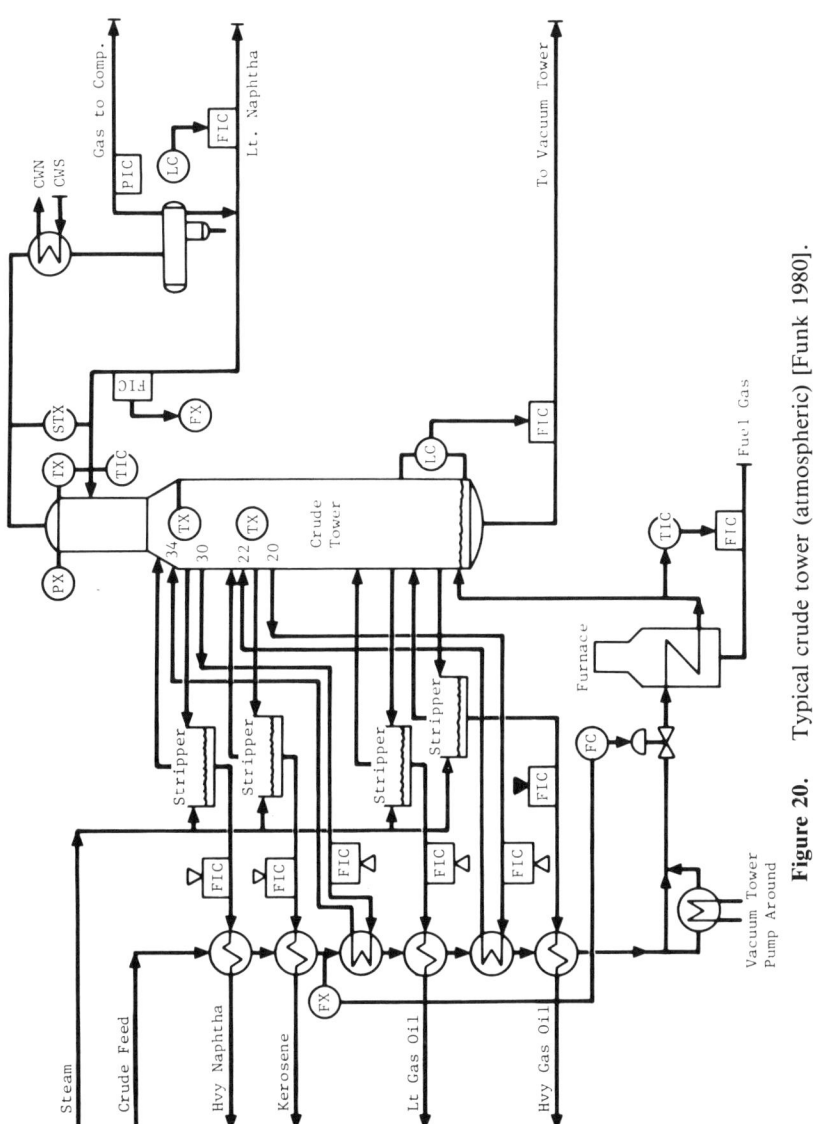

Figure 20. Typical crude tower (atmospheric) [Funk 1980].

- steam-to-hydrocarbon ratio to the primary reformer;
- hydrogen-to-nitrogen (H/N) ratio in the synthesis loop;
- purge rate from the synthesis loop; and
- temperature of the reformer effluent gas.

The H/N ratio control loop is perhaps the most difficult function to control. This loop is also the most important in stabilizing downstream operations. Control of the steam-to-hydrocarbon ratio in the primary reformer ensures adequate steam flow to prevent coking of the catalyst in the reformer tubes. Primary reformer effluent gas temperature control minimizes downstream disturbances while it conserves fuel.

As shown in Figure 21, purge gas is used as part of the fuel to the primary reformer. The heat value of the purged gas, however, is lower than that of the fuel, so it is necessary to constantly adjust the air-to-fuel ratio before the gas enters the furnace. The advantages of computerized control are:

- The H/N ratio is held within a few hundredths of a setpoint.
- Primary reformer temperature is held within $\pm 0.5°C$.
- Pressure variations in the synthesis loop have been eliminated. Synthesis proceeds at a higher pressure, which increases ammonia production.
- Stabilization of the H/N ratio and loop pressure have made the NG_3 converter operate at very stable bed temperatures, allowing its fine tuning and further enhancing production.
- Stability of the process enhances catalyst life, improves efficiency and produces a better on-stream factor for the plant.
- Operator acceptance is excellent, since it makes their job easier.
- Computer control of this plant has saved enough money to pay for the changeover in six months.
- Ammonia production is increased by about 2.5%.

A U.S. Department of Energy (DOE) study [DOE 1979] compared the payback periods in 1962 and 1979 for a process computer system in an ammonia plant (Table XVII). The escalation in utilities and raw material costs should be contrasted with the reduction in the cost of computing power in the 15-yr period. The potential savings increased in 1977 and the payback period has been reduced to a very favorable two months.

The DOE study analyzed payback period, and energy and raw material savings for an ammonia plant under two sets of market conditions: (1) market-limited, and (2) production-limited. When a process is market-limited, the market is saturated and will not accept increased output; thus, increasing production will not result in a greater profit. In the production-limited case, the market will accept all the product that can

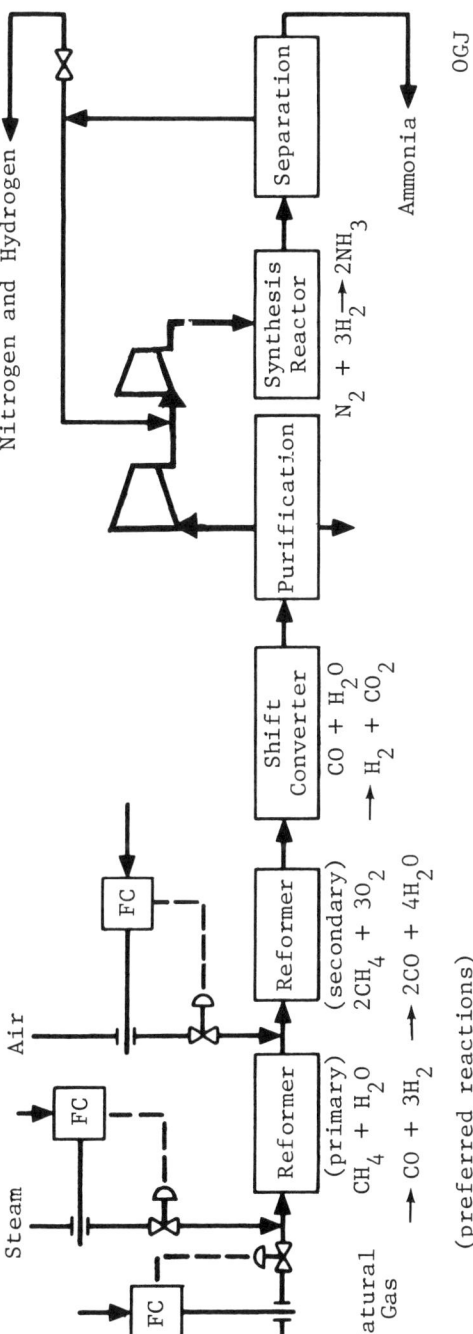

Figure 21.   Ammonia process (Cominco American Inc.).

Table XVII.   Costs and Payback for a Process Computer
System in an Ammonia Plant [DOE 1979]

Characteristic	1962[a]	1977
Capacity (ton/day)	525	525
Costs		
Utilities ($/yr)	600,000	3,120,000
Raw materials ($/yr)	1,220,000	5,462,000
Total operating expenses ($/yr)	1,820,000	8,582,000
Selling price	80	140
Computer investment ($)	450,000[b]	134,500
5% savings in utilities and raw materials per unit of product (total energy savings) ($/yr)	30,000	444,000
3.5% increase in production ($/yr)	500,000	875,000
Total savings ($/yr)	530,000	1,319,000
Payback (months)	20.4	2

[a]Data from Eliot, T. Q., and D. R. Longmire, "Dollar Incentives for Computer Control,"
Chem. Eng. 69:99–104 (1962).
[b]This computer system would cost approximately $1.4 million today, with an inflation rate
of 8%/year.

be made; thus the only limitations are production capacity or throughput
and the ability of downstream processing units to absorb the increased
throughputs.

These two conditions are illustrated in Figure 22, which uses a
1000-ton/day (907-metric-ton/day) ammonia plant as an example. Raw-
material (methane) and utility costs from 1977 were used. The after-tax
payback for a $400,000 process computer system was calculated for a
range of expected savings resulting from reductions in total energy consumption (utilities and raw materials) and increased production. The
market-limited case is shown at a 0% increase in production, and the
production-limited case is shown at 0% reductions in energy and raw
materials, with other variables as follows:

1. market-limited: 0% increase in production, 2% reduction in energy
   and raw materials, and 1.35-yr payback.
2. production-limited: 2% increased production, 0% reduction in energy
   and raw materials, and 0.85-yr payback.

The production-limited case usually offers the faster payback of the two,
since product value is several times greater than raw materials and utility
costs.

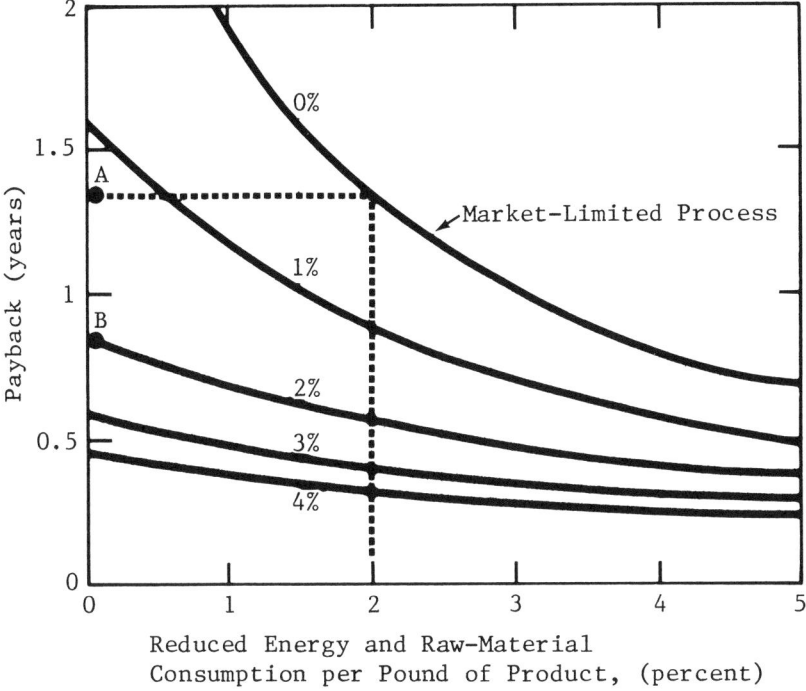

Figure 22.   Savings that result from automation. Payback and energy and raw material savings in a 1000-ton/day ammonia plant resulting from implementation of a $400,000 computer system for process control. Curves represent percent increased production [DOE 1979].

Distillation Systems

Figure 23 shows a schematic flowsheet diagram of a distillation tower with a vapor feed. Vapor overhead products are taken off of a partial condenser and the reflux is pumped to the top tray from the reflux drum. Bottom product is recycled to upstream equipment from a small surge tank. In normal operation the column is governed by upstream constraints that determine both the feed rate and bottom draw. In addition, bounded operation also results from a multicomponent, heterogeneous azeotrope that can form in the column overhead, and by-product purity.

The optimum way of operating the column obviously would avoid the azeotropic conditions and maintain the required product quality. Automatic process control achieves these goals. Figure 24 presents a schematic flow diagram of a distillation tower under computer control, which has the following advantages:

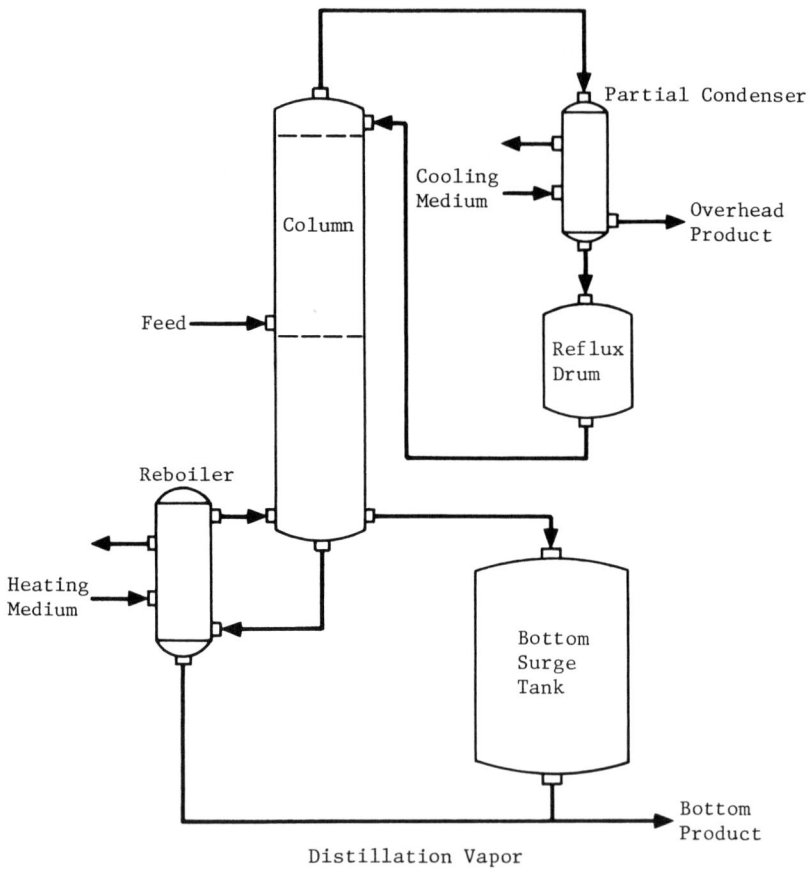

Figure 23. Schematic diagram of distillation tower with vapor feed.

- Column operation is stabilized.
- Automated control operates the column outside the azeotropic region, thus eliminating the need for recycling large amounts of unseparated material.
- There is a smooth transition when there is a change of conditions, so a steady output is maintained.
- Operating near the critical parameters increases the product yield.

An interesting proposal is the automation of a multiple-effect evaporator. Figure 25 provides a diagram of a three-effect feedforward evaporator under supervisory control. Table XVIII lists some of the evaporator's performance characteristics — under conventional computer control and

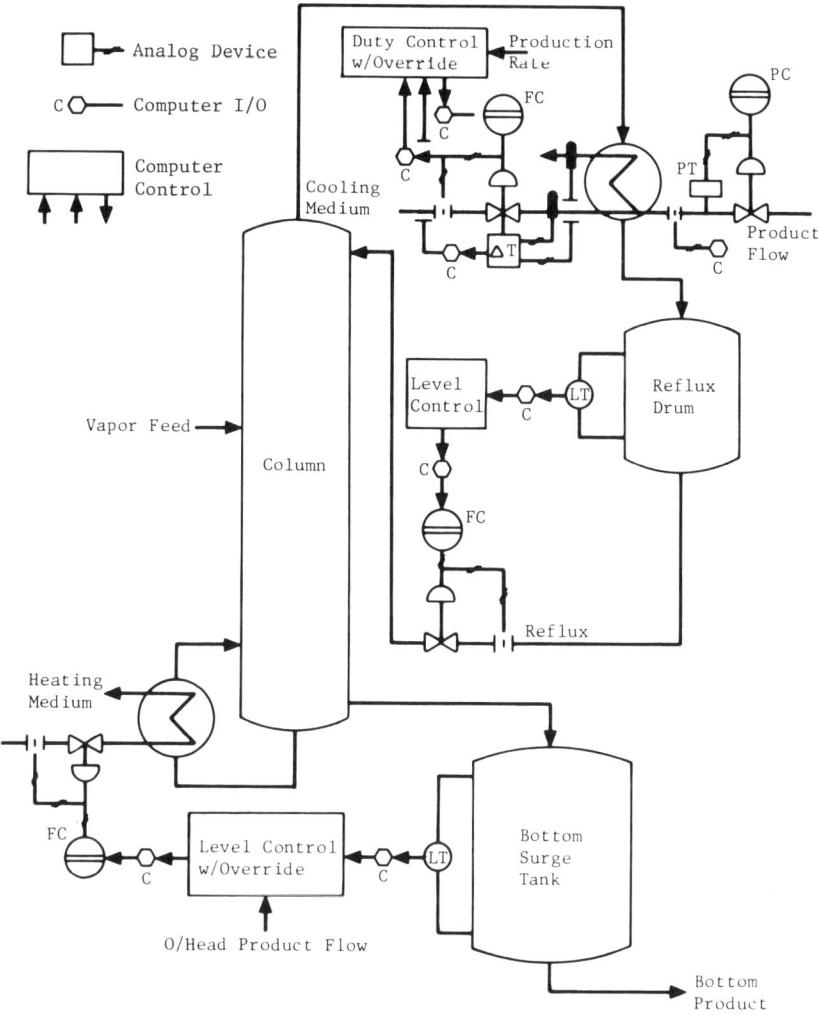

Figure 24. Automatic process control of a distillation tower.

under the proposed supervisory computer control option. The present operating philosophy yields an average product with 43% solids. With the proposed new supervisory control option, the product concentration would be reduced, resulting in a substantial savings of salable product. The new system would yield a 37% solid product — for a greatly improved performance.

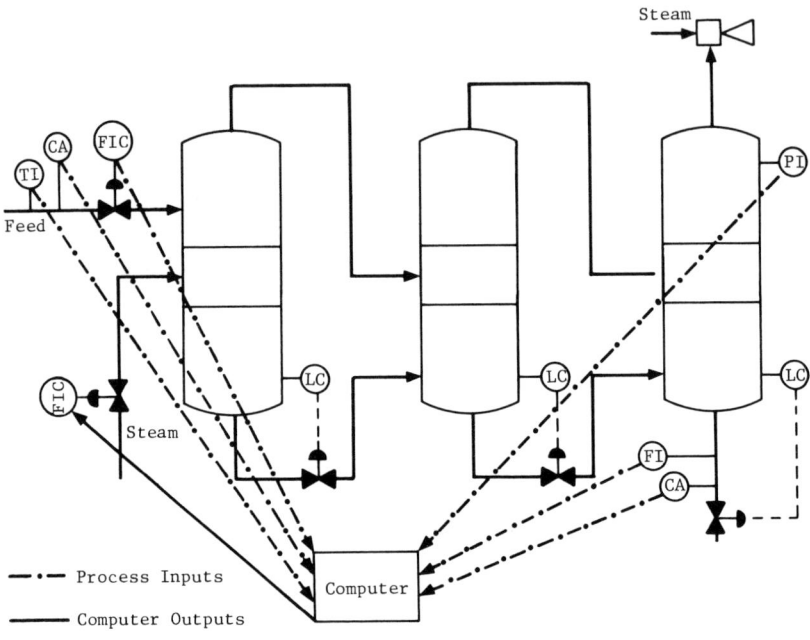

Figure 25.   Supervisory computer control of a multiple-effect evaporator with feedforward [DOE 1979].

Table XVIII.   Multiple-Effect Evaporator[a] [DOE 1979]

	Present Operating Conditions	Predicted with Computer Control
Product Composition (% solids by weight)	43	37[b]
Steam Flowrate (lb/hr)	15,250	14,050
Product Flowrate (lb/hr)	24,530	28,450
Operating Time (day/yr)	120	
Steam Savings at $3.30/1000 lb ($/hr)		3.96
Product Savings ($/hr)		26.80
Total Yearly Savings ($)		88,600

[a]Steady-state feed conditions: flowrate = 83,000 lb/hr; temperature = 190°F; composition = 12% solids by weight.
[b]Minimum allowable product composition = 36% solids by weight.

Such a system would have a payback period of barely more than 1.5 yr, as indicated in Table XIX, which breaks down the estimated cost of a supervisory control system.

Refineries

Digital computers in the refinery segment improve production efficiency and raise quality through more precise control of the production process. Other benefits include better technical and operating data and improved plant safety.

In process control, digital computers are applied to various refining

Table XIX. **Estimated Costs for a Process Computer System for a Multiple-Effect Evaporator [DOE 1979]**

Computer Costs (microcomputer) ($)	
8-Bit processor board	250
Cabinet	200
Power supply	150
16K RAM board	1,800
2K PROM board	120
I/O interface	100
Card rack	200
CRT	1,500
Dual floppy disk (1.5 Mbytes)	2,000
Electrostatic printer	3,000
Analog inputs/outputs (16/4)	3,500
A/D and D/A converters and multiplexers	400
Software	4,000
Total	19,220
Instrumentation Costs ($)	
Three flow transmitters	1,650
Pressure transmitter	850
Current-to-pressure converter	350
Cascade secondary two-mode controller	865
Temperature transmitter	500
Two analyzers (feed and product)	20,000
Total	24,215
Development and Interfacing	30,000
Total costs ($)	73,435
Project life (yr)	10
Payback period (yr)	1.6

processes, ranging from crude distillation to on-line gasoline blending. Computerized installations can be found in 25% of all refineries, accounting for more than two-thirds of the industry's crude capacity. It is expected that practically all future refineries will incorporate one or more process control digital computer systems. Open-loop control is most common; however, closed-loop control is used increasingly in the newest and largest installations. The trend is toward use of mini- or microcomputers that control separate functions while linked to a central control center.

In a refinery visited by the U.S. Department of Labor [DOL 1979], the computer monitored information from more than 1000 electronic instruments, such as chromatographs, mass spectrometers and octane analyzers, that were located at the process unit and continuously measured product quality. Such a system is important because of the speed with which problems can be corrected. All U.S. refineries use analyzers, but the number and sophistication of the instruments vary with the size and complexity of the plant.

Crude units have a built-in need for interactive control. Because of the large volume of energy and materials involved, enormous savings are possible with small changes in operating conditions affecting yields.

Several field installations have been reported as under closed-loop control [DiBiano 1981]. An example is the hybrid process control microprocessor-minicomputer, installed on a crude unit at the Phillips' Sweeney, Texas, refinery.

Catalyst Plant

Catalysts are expensive and essential to many industrial chemical processes. Union Carbide uses process control in the latter stages of catalyst development to eliminate wasteful expenditure of reactants, catalyst and energy [Gregory and Young 1979].

The use of computer control to stabilize individual operations has been demonstrated at the El Dorado, Arkansas, nitric acid plan of Monsanto Company [Wright and Powers 1980]. The control system has made it possible to reduce ammonia consumption (the feed) by 1.4% and to lower the rate of expensive catalyst loss by 21%, for a fixed volume of nitric acid production.

Specialty Chemicals

Union Carbide has installed a digital control system to consolidate its specialty chemicals fabrication at its South Charleston, South Carolina, plant [Thurston 1979]. The central computer, through its associated sensors, measures and controls flow, pressure and temperature in two manufacturing areas. The unit operations involved include oxidation, furnace cracking, dehydration, distillation, reaction, evaporation and refrigeration.

The observed benefits relative to previous control systems include:

- reduction in raw materials costs;
- operation of critical units in a semicontinuous process twice as long as previously possible;
- improved ability to build an inventory of finished products;
- important energy savings; and
- practical potential for optimizing the production of specialty chemicals.

Instrumentation

Pressure, level, other parameters, flow and temperature (PLOFT) are monitored in almost all industrial processes [Bailey 1980; *Chemical Week* 1980a,b]. The trend is to put these instruments on-line with the data storage and reduction capability. Favorite instruments being automated include gas chromatographs, pH analyzers and IR spectrometers.

Several arrangements are possible:

- microprocessors within the analytical instrument;
- external microprocessor governing several instruments;
- connection of instruments to a control processor; or
- microprocessor control of PLOFT instrumentation organized into a plant control complex.

On-stream analyzer shipments are expected to grow 15% per year, reaching $465 million in 1985 and $895 million in 1990. Analyzer dedicated data reduction systems are expected to show 14% real growth rising from $109 million in 1978 to $194 million in 1985.

GLASS

Industry Characteristics

The glass products industry includes four SIC code categories as given by DOC: flat glass (SIC 3211), glass containers (SIC 3221), pressed and blown glass (SIC 3223), and products of purchased glass (SIC 3231). The value of glass products shipments in 1977 totaled about $9.2 billion, with that amount divided between the four categories as follows:

- flat glass: $1.5 billion
- glass containers: $3.66 billion
- pressed and blown glass: $2.12 billion
- products of purchased glass: $1.84 billion

Energy purchased in 1976 by the glass manufacturing industry amounted to 285.7 × 10^{12} Btu (3.01 × 10^{17} J) (Table XX). The primary fuel used in the industry is natural gas, followed by oil, electricity, and butane or propane (Table XXI). Most of the energy is consumed in the melting process.

This can be shown in the following breakdown of energy consumption per ton of product in glass container manufacturing.

Operation	Energy Consumed	Percentage
Melting	7.67	70
Refining	1.66	15
Framing	0.6	5
Post Framing	1.11	10
Total	11.04	100

To orient the reader, this section provides process flow diagrams for flat glass (Figure 26), glass containers (Figure 27), pressed and blown glass (Figure 28), and products of purchased glass (Figure 29). Each figure shows unit operations from batch handling, melting fining, forming and post forming to product handling. In reviewing these illustrations, the reader should note the large temperature variation in each function. This is a key energy problem in the glass industry.

The glass industry has made substantial attempts at energy conservation and has been quite successful, as can be seen in Table XXII. Over

Table XX.   Energy Consumption in Glass Manufacturing
[ORAU 1980]

	1976 Consumption of Purchased Fuels and Electricity (10^{12} Btu)	1976 Value of Shipment (10^6 Dollars)	10^6 Btu/10^3 of Shipment
Flat Glass	53.5	1,006.3	53.2
Glass Containers	147.2	3,047.0	48.3
Pressed and Blown Glass, NEC[a]	67.7	1,505.6	45.0
Product of Purchased Glass	17.5	1,336.1	12.9

[a]NEC = not elsewhere classified.

Table XXI.   Trends in Energy Efficiency for Three Segments
of the Glass Industry [Stewart 1978]

	1972	1973	1974	1975	1976	First Half 1977
Flat Glass						
Production (10^6 tons)	2.59			2.22	2.91	1.49
Btu (10^{12})	53.1			39.9	47.8	24.1
Ratio (10^6 Btu/ton)	20.5			18.0	16.4	16.2
Percentage improvement[a]				12.3	19.9	21.3
Glass Packaging						
Production (10^6 tons)	11.5	11.8	11.9	12.1	12.7	6.4
Btu (10^{12})	133.3	135.0	135.0	135.1	137.7	66.8
Ratio (10^6 Btu/ton)	11.59	11.44	11.34	11.17	10.84	10.49
Percentage improvement[a]		1.3	2.2	3.6	6.5	9.5
Pressed and Blown Glass						
Production (10^6 tons)	1.81		1.87	1.68	1.95	1.13
Btu (10^{12})	61.9		59.2	51.2	57.1	30.7
Ratio (10^6 Btu/ton)	34.2		31.7	30.5	28.3	27.2
Percentage improvement[a]			7.4	10.9	14.4	20.8

[a]Percentage improvement is calculated relative to 1972, the base year.

five years, energy consumption has been cut by 21% in the flat, and pressed and blown glass segments and by 10% in glass packaging.

Process control and automation in the glass industry have been applied to batch plant control, glass melting control, process control in the manufacture of glass containers, process control for the flat glass industry, inspection control and production recording [Edgington 1979a,b], in order to:

• reduce cost;
• improve quality;
• increase productivity;
• provide effective management information;
• reduce dependence on manual control; and
• improve working conditions.

Glass Melting

As early as 1964, one manufacturer demonstrated the use of direct digital control of a glass melting process. The benefits included lower

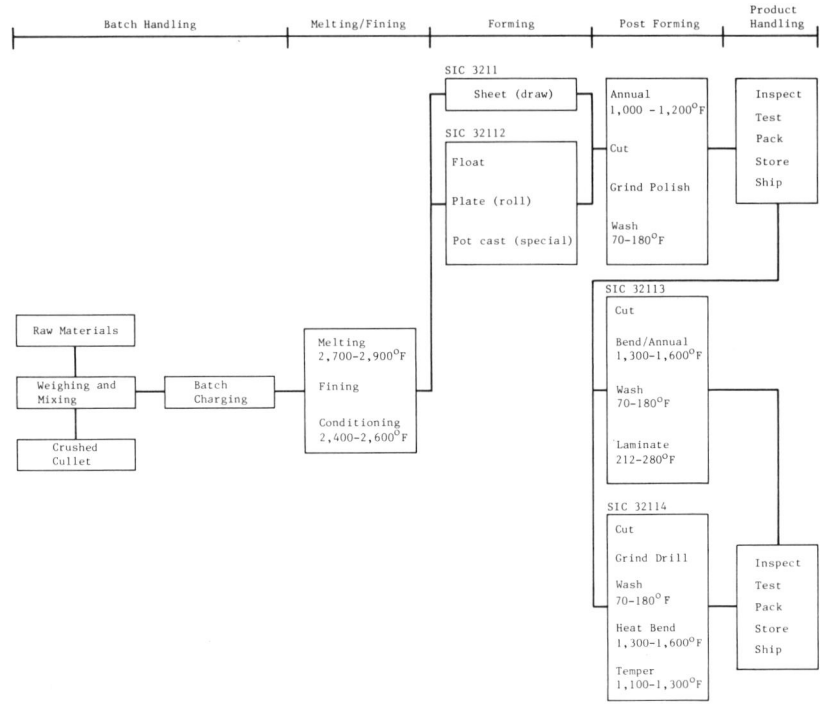

Figure 26.   Process flow diagram: SIC 3211, flat glass [Battelle 1977].

fuel usage, better temperature control and improved product quality. Despite this early success, the industry has not hurried to automate for a variety of reasons, including the high installation cost of digital systems.

Today, the situation is changing rapidly. High fuel costs, shortages of qualified manpower, pollution control requirements, a high furnace rate requirement and the advent of small, reliable, low-cost computers all indicate the need for renewed interest in automation.

One of the most important variables in the glass industry is the temperature at which the raw materials are processed. At low temperature, unmelted raw material (commonly called batch stones) will be drawn through the melter, resulting in quality problems in the finished product. In practice, the melter operating temperature is kept at levels considerably higher than the critical temperature (batch stone temperature), so that disturbances within the melter do not result in temperatures below the critical point. In many instances the difference between the operating temperature and the critical temperature is about 200°F (93°C). This higher operating temperature results in extra fuel use, while reducing the furnace's useful life.

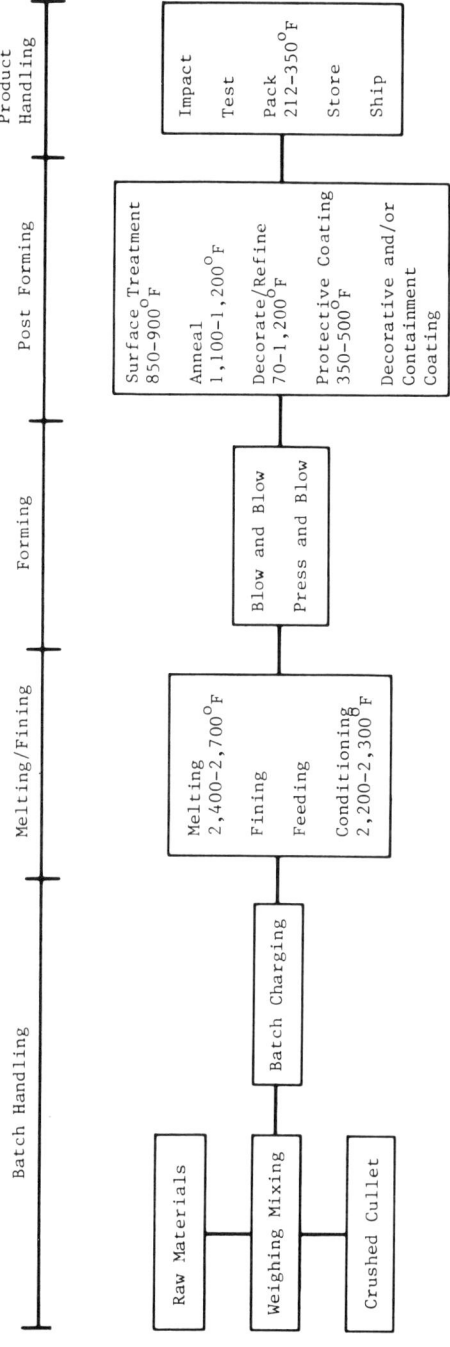

Figure 27.   Process flow diagram: SIC 3221, glass containers [Battelle 1977].

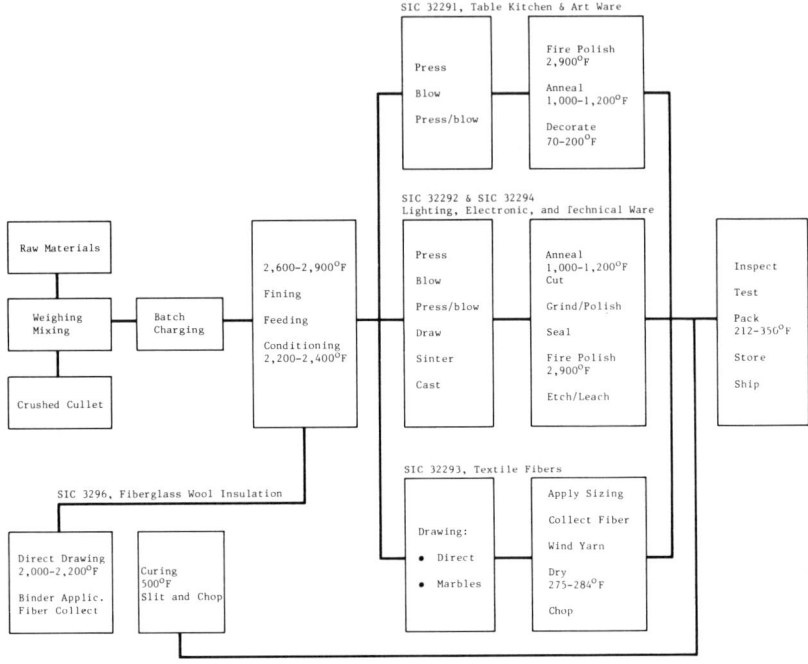

Figure 28.   Process flow diagram: SIC 3229, pressed and blown glass, NEC, and SIC 3296, fiberglass wood insulation [Battelle 1977].

Figure 30 is a schematic flow diagram of a computerized melter refiner and forehearth process layout [Finger 1978]. Figure 31 is an analog representation of the digital control technique for the melting temperature control. The computerized system controls:

- melter glass temperature;
- melter crown temperature;
- melter combustion airflow;
- melter pressure;
- melter glass level;
- refiner glass temperature;
- refiner crown temperature;
- forehearth glass temperature; and
- forehearth crown temperature.

In addition, the system monitors:

- glass bushing temperatures; and
- spool winder operations.

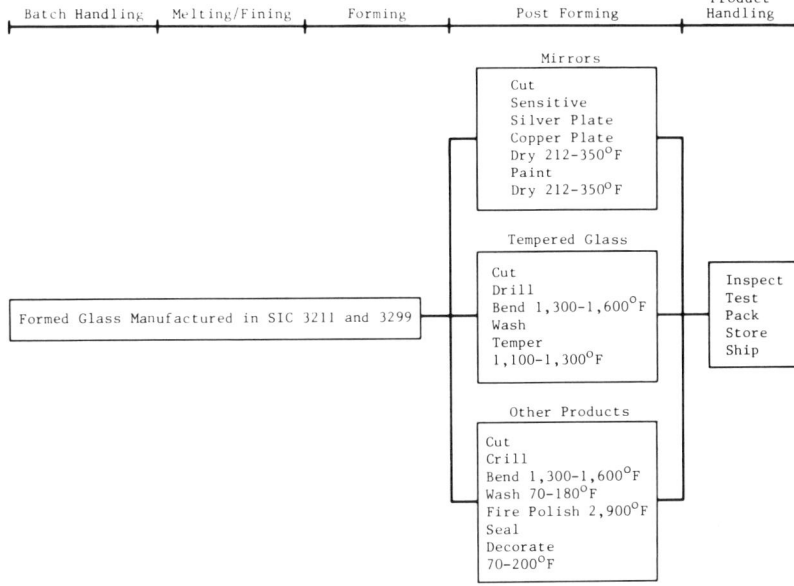

Figure 29.   Process flow diagram: SIC 3231, products of purchased glass [Battelle 1977].

Several operating advantages are readily obtained:

- reduced energy consumption;
- more accurate temperature control throughout the process;
- longer furnace life;
- better coordination of melter, refiner and forehearth control; and
- improved operating control.

Glass Forming and Tempering

The gas hearth is used to shape and temper glass plates for automobiles [Loibl 1978]. Achieving desired glass properties requires maintaining a specific temperature and pressure profiles inside the Lehr.

To monitor and control these critical process variables, a digital system has been installed at the glass division of the Ford Motor Company in Dearborn, Michigan. The digital computer allows:

- presentation of information to the operator in a readily understandable digital form;

Table XXII.　Energy Use in Glass Manufacturing
[Battelle 1975]

Process Step	Unit	Units per Net Ton of Product	10^6 Btu per Unit	10^6 Btu per Net Ton of Product
Melting				
Natural Gas	10^3 ft^3	5.95	1.000	5.95
Electricity	kWh	64.52	0.0034	0.22
Distillate Oil	gal	7.35	0.139	1.02
Residual Oil	gal	3.03	0.150	0.45
Propane/Butane	gal	0.35	0.0955	0.03
Total				7.6
Refining and Conditioning				
Natural Gas	10^3 ft^3	1.64	1.000	1.64
Electricity	kWh	4.98	0.0034	0.02
Total				1.7
Forming				
Natural Gas	10^3 ft^3	0.23	1.000	0.23
Electricity	kWh	107.85	0.0034	0.37
Propane/Butane	gal	0.0	0.0955	0.00
Total				0.6
Post Forming				
Natural Gas	10^3 ft^3	1.01	1.000	1.00
Electricity	kWh	25.63	0.0034	0.09
Propane/Butane	gal	0.17	0.0955	0.02
Total				1.1
Grand Total				11.0

- collation of data and preparation of process status reports;
- collection and storage of process information;
- easy modification of cascade control strategies;
- implementation and modification of supervisory control strategies; and
- preparation of production reports.

The computer controls the Lehr temperatures and quench pressure and monitors plenum pressures and glass temperatures. Lehr status data are used to analyze the operation of the gas hearth.

Cold End Automation

A variety of automatic systems have been applied to warehousing in the glass industry, which encompasses all steps after the finished ribbon

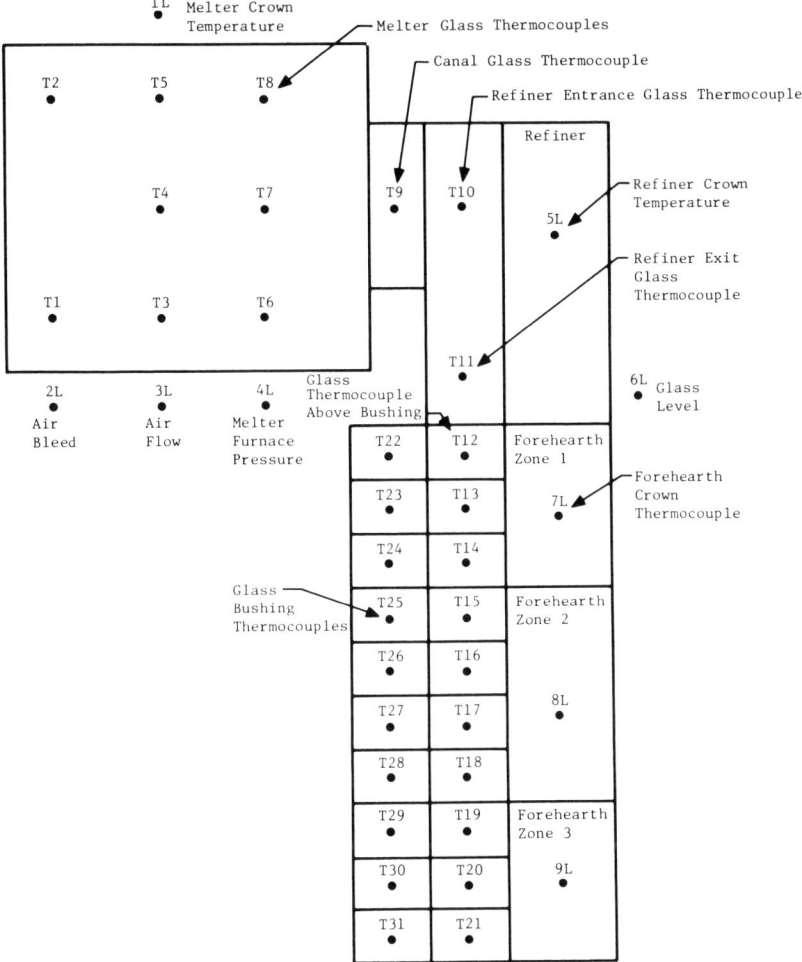

Figure 30. Automated glass melter-refiner and forehearth process layout. T = glass thermocouples master control loops; L = primary control loops. [Finger 1978].

of glass leaves the annealing Lehr and before the cut pieces are distributed to customers [Branch 1976]. The essential steps (Figure 32) are:

1. Inspection: float ribbons are produced too quickly for human inspection; hence, automatic inspection has been developed.
2. Cutting: mechanical and electrical devices are used to drive a cutting head across the ribbon to produce the cross-cut score that, when snapped, produces the desired plate.

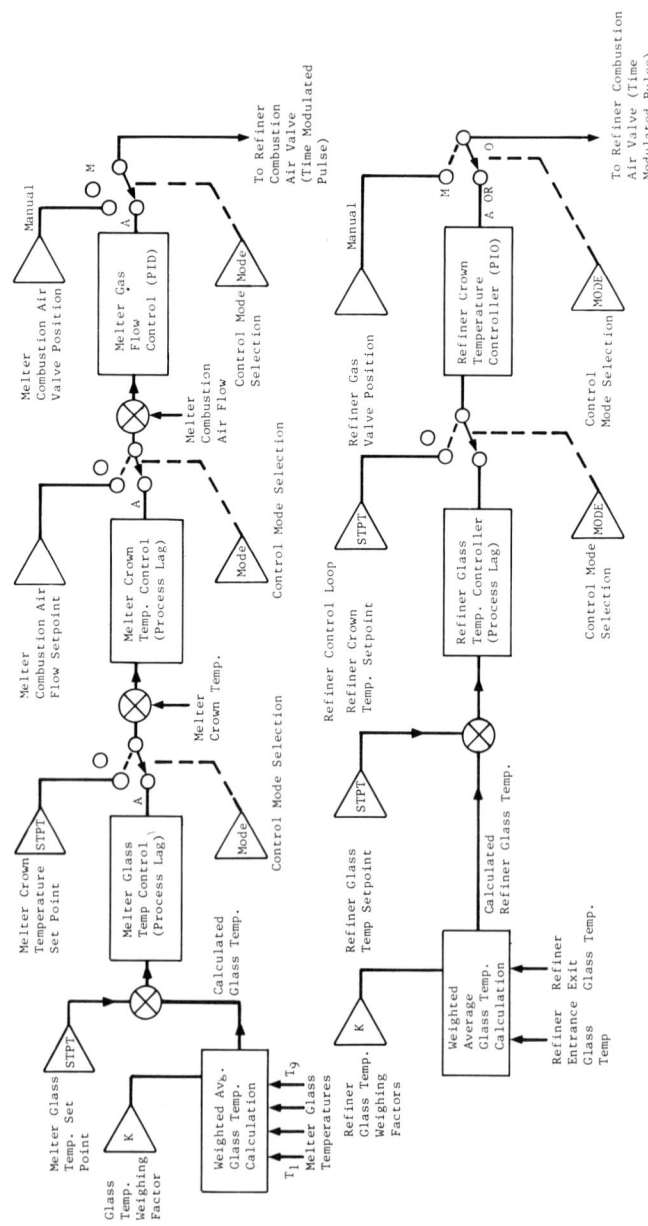

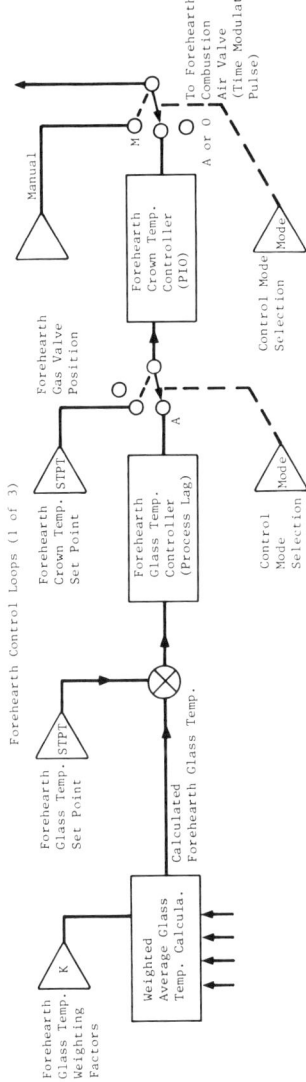

Figure 31.   Analog representation of the digital control technique for melter temperature control. M = manual control of final control elements; 0 = operator input of setpoint; A = automatic calculation of setpoint [Finger 1978].

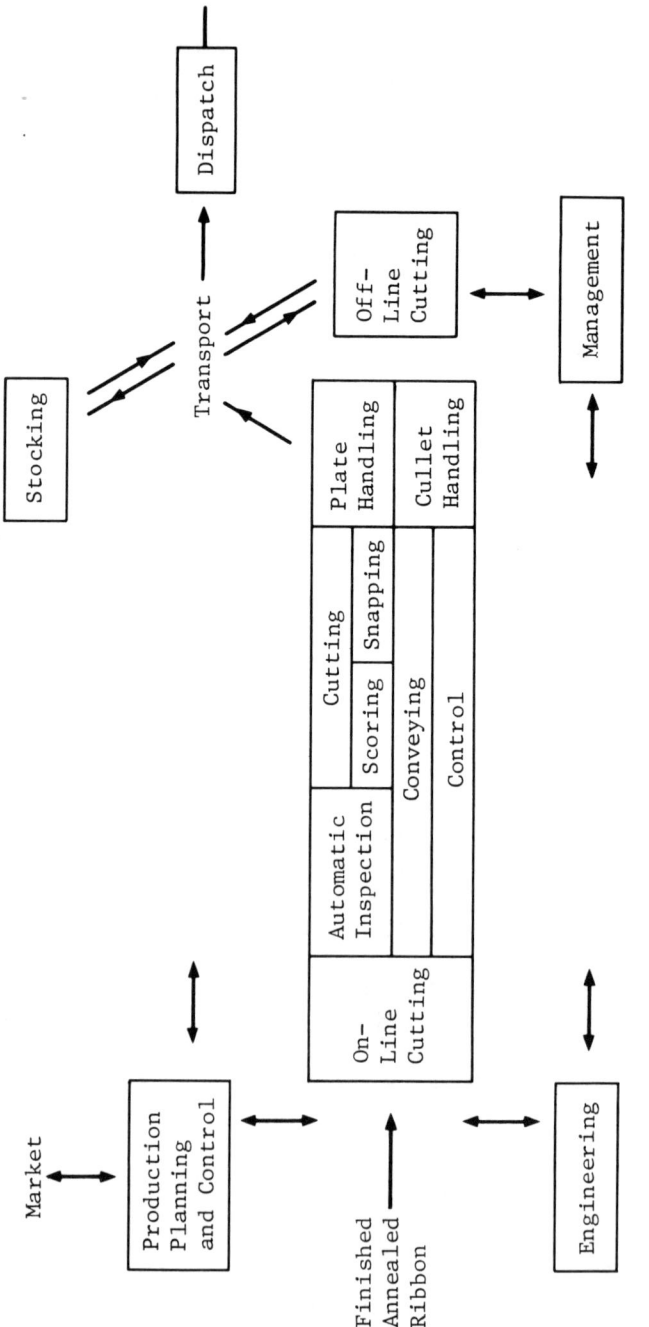

Figure 32.   Schematic of warehouse system [Branch 1976].

3. Handling: plates are removed from the line by automatic cranes that transfer them to the stack being built on a tilting table.
4. Conveying: matched speed conveying is essential to avoid scarring.

PHARMACEUTICAL PRODUCTS

Industry Characteristics

Like the glass industry, the pharmaceutical industry can be divided into SIC categories: biological products (SIC 2831), medicinals and botanicals (SIC 2833), and pharmaceutical preparations (SIC 2834).

Biological products consist of bacterial and viral vaccine, toxoids, serums, plasmas and other blood derivatives for human or veterinary use. The category has been expanding in recent years, with the total value of shipments increasing to $899 million in 1977 — an increase of 150% over 1972. In 1977 the industry had 15,700 employees — 55% more than in 1972.

Medicinals and botanicals comprise bulk organic and inorganic chemicals and their derivatives, as well as bulk botanical drugs and herbs. This category also includes natural-origin endocrine products; basic vitamins; and active medicinal compounds, such as alkaloids, from botanical drugs and herbs. The value of shipments classified in this category has risen sharply — reaching $1.89 billion in 1977 — an increase of 271% over 1972. In 1977 the industry had 14,400 employees, for an increase of 85% over 1972.

Pharmaceutical preparations comprise drugs in pharmaceutical preparations for human or veterinary use. Products of this industry consist of two important lines: (1) pharmaceutical preparations sold largely to the dental, medicinal or veterinary professions; and (2) pharmaceutical preparations sold to the public. The value of shipments for establishments classified in this category amounted to $11.6 billion in 1977 — an increase of 60% over 1972. In 1977 the industry had 126,400 employees, an increase of 13% over 1972.

Automation

In the pharmaceutical industry, computers are used on-line to control chemical processes, pharmaceutical operations, inventories and warehouses, as well as analytical instruments in laboratory, physiological and medical tests. Merck is an example of one drug company using automation extensively [Thome et al. 1979].

The pharmaceutical industry is a batch-processing industry. This results in reduced use of equipment, and more costly processing delays and waste due to faulty equipment scheduling and product contamination. As with any other manufacturing activity, automation permits the operator to monitor, regulate and sequence the process, and to learn the status of the different parts of the process. In batch processing, this takes on an added importance because it permits idle equipment to be used as soon as it is vacated. The advantages are:

- cost-effective sequencing;
- improved overall plant efficiency;
- reduced plant energy consumption; and
- improved efficiency of chemical, pharmaceutical and fermentation operations.

One process that has been identified as a good candidate for automation using small computers is fermentation [Bungay 1980], which is used extensively in the pharmaceutical industry. Temperature, pH and turbidity are commonly monitored.

Other processes would also be well suited to automation, and the Food and Drug Administration (FDA) predicts a widespread move toward fully automated manufacturing procedures in the pharmaceutical industry [*FDC Reports* 1981]. Computers will be programmed to weigh ingredients, control the mixing and granulation process, select the dosage, and control the packaging of the final product. An example of such a fully automated production facility can be found at the Merck facility in Elkton, Virginia [Wheeler 1977], and the Smith Kline Corporation Gelatine Manufacturing Facility in Philadelphia, Pennsylvania [Martyn 1974].

FOODS AND BEVERAGES

Industry Characteristics

The U.S. food processing industry generated an estimated $227.6 billion in shipments in 1979—a 7.5% increase over 1978. In 1976, the food and beverage manufacturing group used 937.5×10^{12} Btu (1.54×10^{15} J) of energy. Table XXIII presents the industry's consumption of purchased fuels and electricity. The variations in energy consumption per unit of product within an industry are due to plant size, plant design (efficiency), type of process used and the age of boilers. Most of the

Table XXIII.   Food and Kindred Products: Energy Use
[DOC 1976-1978]

Priority Group	SIC Code	Industry	Purchased Fuel and Electrical Energy (10^{12} Btu)
I	2063	Beet sugar	92.7
	2046	Wet corn milling	83.1
	2011	Meat packing plants	64.8
	2082	Malt beverages	46.8
	2075	Soybean oil mills	46.7
	2033	Canned fruits and vegetables	45.2
	2051	Bread, cake and related products	40.6
	2026	Fluid milk	39.0
	2062	Cane sugar	33.7
	2037	Frozen fruits and vegetables	30.1
	2099	Food preparation, NEC[a]	29.2
	2048	Prepared feeds, NEC	26.7
II	2079	Shortening and cooking oils	25.8
	2077	Animal and marine fats and oils	23.6
	2022	Cheese, natural and processed	22.4
	2086	Bottled and canned soft drinks	20.9
	2023	Condensed and evaporated milk	20.3
	2032	Canned specialties	20.0
	2016	Poultry dressing plants	17.7
	2013	Sausage and prepared meats	17.4
	2085	Distilled liquor	15.0
	2048	Pet food	13.9
	2065	Confectionary products	13.8
	2034	Dehydrated food products	12.7
	2061	Raw cane sugar	12.5
	2038	Frozen specialties	11.0
	2041	Flour and other grain mill products	10.8
	2052	Cookies and crackers	10.8
III	2095	Roasted coffee	10.4
	2083	Malt	9.4
	2043	Cereal preparations	9.4
	2074	Cottonseed oil mills	6.6
	2024	Ice cream and frozen desserts	6.2
	2035	Pickles, dressings and sauces	5.8
	2021	Creamery butter	4.6
	2066	Chocolate and cocoa products	4.5
	2087	Flavoring extracts and syrups, NEC	4.4
	2084	Wines, brandy	4.2
	2092	Fresh or frozen packaged fish	3.8
	2017	Poultry and egg processing	3.7
	2044	Rice milling	3.5

Table XXIII,   continued

Priority Group	SIC Code	Industry	Purchased Fuel and Electrical Energy (10¹² Btu)
	2091	Canned and cured seafood	3.1
	2076	Vegetable oil mills, NEC	2.5
	2097	Manufactured ice	2.4
	2098	Macaroni and spaghetti	2.4
	2045	Blended and prepared flour	1.9
	2067	Chewing gum	1.3
Total			937.5

a NEC = not elsewhere classified.

fossil fuels purchased by the food industry are used for process heat; a small percentage is used for space heating, and in some cases, refrigeration and onsite electric power generation [Vitullo undated]. Examples of energy consumption by process appear in Table XXIV.

The industry uses automation and computer control to cope with increasing labor and energy costs. Energy savings strictly attributable to automation and computer-controlled processes were not found in the literature; however, most segments of the industry reported significant improvements to the DOE Voluntary Business Energy Conservation Program (Table XXV). More efficient use of facilities is cited as one contributing factor.

Other benefits attributed to the use of automation and computer controlled processes include:

- product of a more uniform quality;
- reduction of off-grade products;
- increased yields;
- reduced requirements for tanks, pumps and space; and
- less analytical laboratory support.

Automation systems for complex food and beverage plants fall into four categories: (1) mechanization and remote control, (2) analog control, (3) hard-wired systems, and (4) programmable systems. Selected examples of the use of automation are discussed for the dairy and baking sectors.

Table XXIV.   Energy Distribution Percentage Use by Process [DPRA 1976]

Use Category	Meat Processing Plant, SIC 2011	Sausage and Prepared Meats, SIC 2013	Poultry Dressing Plants, SIC 2016	Poultry and Egg Processing, SIC 2017
Direct Fuel Use				
Space heat	1.2	0.6	2.5	1.2
Singeing	1.6			
Smoke houses	2.4			
Afterburners	1.6	3.3		0.3
Process ovens		5.0		0.5
Hot water	1.2		0.5	
Miscellaneous		2.2	2.0	1.0
Fuel and Boiler Losses	19.8	19.2	19.0	17.4
Steam				
Rendering	19.8	25.9		22.9
Process	9.1	14.7		20.1
Water	28.1		50.4	
Smoke houses		4.4		
Other		2.8	3.7	6.7
Electricity				
Lights	2.0		1.1	2.4
Heating, ventilating and air-conditioning	0.6	0.3	0.9	0.6
Mechanical power	5.7	9.9	6.8	15.4
Refrigeration	5.2	5.1	13.2	9.0
Primary waste treatment	0.6			0.6

Table XXV.   Percent Energy Savings per Unit Output:
1978 over 1972 Base Year [DOC 1980a-f]

Association	Actual	1980 Goal
American Meat Institute	27	12
National Food Processors	22	11
Brewing	22	8
National Soft Drink	22	14
Baking	− 2.6	14

Milk and Milk Products

Productivity in the fluid milk industry increased at an average annual rate of 4% between 1958 and 1977, as measured by output per employee-hour. One of the major factors was a trend toward fewer and larger plants using technology that improved milk-processing operations. Computers have been used in the standardizing operation to monitor and control skim milk flow and determine the butterfat content of different products. The introduction of better, faster filling machinery has also contributed to productivity. Further gains were realized by linking improved filling machinery with automatic casing and stacking equipment. In a plant where case-handling equipment is used, one person completes a truck loading operation in less than 30 min. Before installation of the equipment, the same task required three workers and close to an hour [Persigehl and York 1979].

Milk is piped to various processing plants for pasteurization, cream separation, homogenization and other processing. Drinking milk is pumped to a commercial milk tank, a drinking milk tank or a drinking-chocolate tank; cream goes to a cream storage tank. Measured quantities of bacteria cultures are added. Tanks, containers, pumps and stirrers are cleaned at regular intervals using cleaning-in-place (CIP) procedures. A diagram of material and energy flow for fluid milk plants is presented in Figure 33.

In the automated system installed at Krejelder Milchof-Krefeld, West Germany, the plant is controlled by one operator, using a diagram displaying the state of all processes. The operator selects receiving and destination tanks, the process for each batch of milk, the CIP program to be applied, and the production quantities per product. The required quantity of milk is automatically fed to each processing plant. Interlocks ensure that once a cleaning cycle is started, it cannot be stopped until completed.

The advantages of the automated system are flexibility, simplified

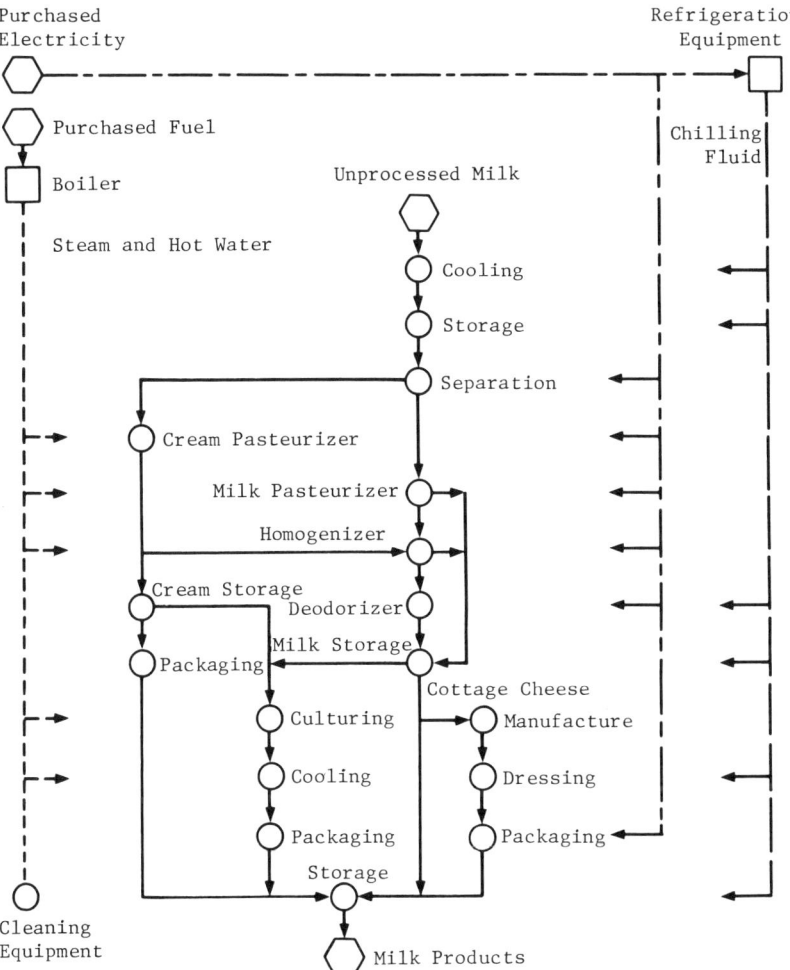

Figure 33.   Material and energy flow for fluid milk plants [DPRA 1976].

operation and control. It has high reliability, minimizes staff requirements and lowers overall cost.

Skim Milk

The production of skim milk, its temporary storage and subsequent
blending with raw milk and condensed skim milk or nonfat milk solids

requires substantial additional equipment and space when handled in the conventional approach involving batch-blending. An alternative is a continuous "on-line" approach incorporating the centrifugal separator in the HTST pasteurizing system [Siebergling 1974]. Standardizing/blending operations are controlled through the use of meters and digital control equipment. The flow ratio is established on the basis of input data, including batch size and composition of raw products and desired finished product. This computer system controls cream discharge from the centrifugal separator to establish the fat test of the cream. A portion of the cream removed is recombined with the resultant skim to produce a standardized product that may be fortified through the addition of a suitable slurry. The system also provides for volumetric control of water/product separation on startup and shutdown.

Increased productive capacity, processing flexibility and accuracy of formulation are among the benefits of the automated system. Automation eliminates the need for batch standardizing and makes possible the handling of a complete day's production on the basis of initial composition. Substantial product savings result from the ability to run a variety of products through the system without intermediate rinses or mixing one product with another.

Cheese

For cheddar cheese production, pasteurized milk is pumped to steam-jacketed cheese vats with the starter culture metered en route. A total of 80% of the milk is converted into curds, 20% into whey. Curds are drained on cheddar masters (perforated slat conveyors). Cheddaring requires pressure on the curd. Granular curd is lifted by air to the top of a cheddaring tower. The weight of the curd as it builds up compresses the lower volumes, which are cut away in blocks at the base. These blocks are chipped and passed to a press, which produces a vertical column of cheese that is in turn cut into blocks.

At the Maelor Creamery in Denbighshire, North Wales, the system provides for a fully automatic routing of milk, complete CIP, control of all equipment (including cheddar masters), display of fault conditions, and control of cheese-vat filling, valves and piping. Program changes can be made by company personnel who have no computer knowledge.

A cheese factory in Edewecht, West Germany, built in 1975 at a cost of 22 million deutsche marks is one of the most highly automated to date. In 1978 it processed about 990,000 lb (450,000 kg) of milk daily into cheddar, gouda and edam cheese. In this plant manpower requirements are low. For example, in the cheesemaking hall (about 27 × 60 yd or

25 × 55 m) no more than two to three workers are needed per shift. A detailed description of the plant's operations is presented in *Milk Industry* [1978].

A continuous, fully automated system for producing mozzarella cheese is available from American Pioneer Corporation. The modular system is capable of making up to 8,000 lb/hr (3,600 kg/hr), in sizes ranging from 8-oz (227-g) chunks to 20-lb (9-kg) blocks. The cheese is popular in the United States (domestic sales in 1979 exceeded $600 million, and cheesemakers who have installed the system claim to achieve paybacks of 7 months to 1 year). The system is reported to automatically cook and curd with no fat loss; and automatically mold, form, cool and route the cheese to brine tanks. These operations take place in 140 ft^2 (13 m^2) of floor space [Frost 1980].

Bakery Products

Technology in the baking industry has not changed substantially over the past 20 years. The limited number of changes that have occurred have been confined to improving existing machinery rather than changing design or functions. In the future, the trend is expected to be toward larger plants and possibly to automated systems for more bread baking plants if they are made economically feasible by higher labor costs.

Advanced technologies are already widely established in the cookie and cracker segment. The bakery products industry employed approximately 228,400 workers in 1978—almost 25% below the number employed in the peak year of 1956. Employment declined steadily as the industry moved to greater concentration and use of mechanized, labor-saving production techniques.

The industry is composed of two major groups: manufacturing of perishable products (bread and cakes); and dry products (biscuits, cookies and crackers). The bread sector is by far the larger, with almost five times the number of employees and three times the value of shipments.

The largest proportion of cookie and cracker shipments comes from high-capacity plants; therefore, the economies resulting from the installation of automated equipment have long been recognized. Bulk handling of material, high-capacity mixing, mechanized processing and automated packaging are common in the industry. Automated technologies were introduced in the bread baking sector in the 1950s. For larger plants, automated processing systems were developed that included liquid fermentation and continuous mixing. However, the bread produced by continuous mixing has not gained consumer acceptance. Liquid fermentation is used almost exclusively with conventional processing rather than with a continuous system. Although no significant advances have occurred in bread baking technology, equipment has been improved to reduce manual processing and handling. Pneumatic conveyors transfer bulk materials to storage and mixing operations. Automatic batching and weighing devices quickly deliver ingredients to mixers. Increasing labor costs may make adoption of more automated machinery feasible in medium and small plants.

The bakery products sector is not energy-intensive per pound of product, but high volume makes it a major energy consumer. In 1976 natural gas supplied 53% of the energy consumed. Schematic material and energy flowcharts for a bread and cake plant are included as Figures 34 and 35, respectively.

Before introduction of an on-line, real-time information system, batch computers were used in mixing biscuit ingredients and in process control.

The floor layout of the oven room (Figure 36) shows the ovens and wrapping and packing machines used at United Biscuits Ltd., Harlesden, England. Various terminals used in the production information system are distributed around the factory floor. Four basic systems encompass all factory applications of the computer:

1. production/process: monitoring and control;
2. people: data, movement and analysis;
3. maintenance: stores control and machine histories; and
4. materials: bulk ingredients monitoring and control, and packaging inventory.

The United Biscuits automated production information system includes four applications in the production/process subsystem:

1. production monitoring;
2. on-line turn-of-scale measurement;
3. production count; and
4. factory floor messages.

The production monitoring application maintains records of stoppages and reasons for stoppage in ovens and wrapping machines. Four production monitoring reports are generated on the basis of stoppage logs:

1. current status summary: available at all terminals, covers all ovens, wrapping machinery and indicates whether they are producing;

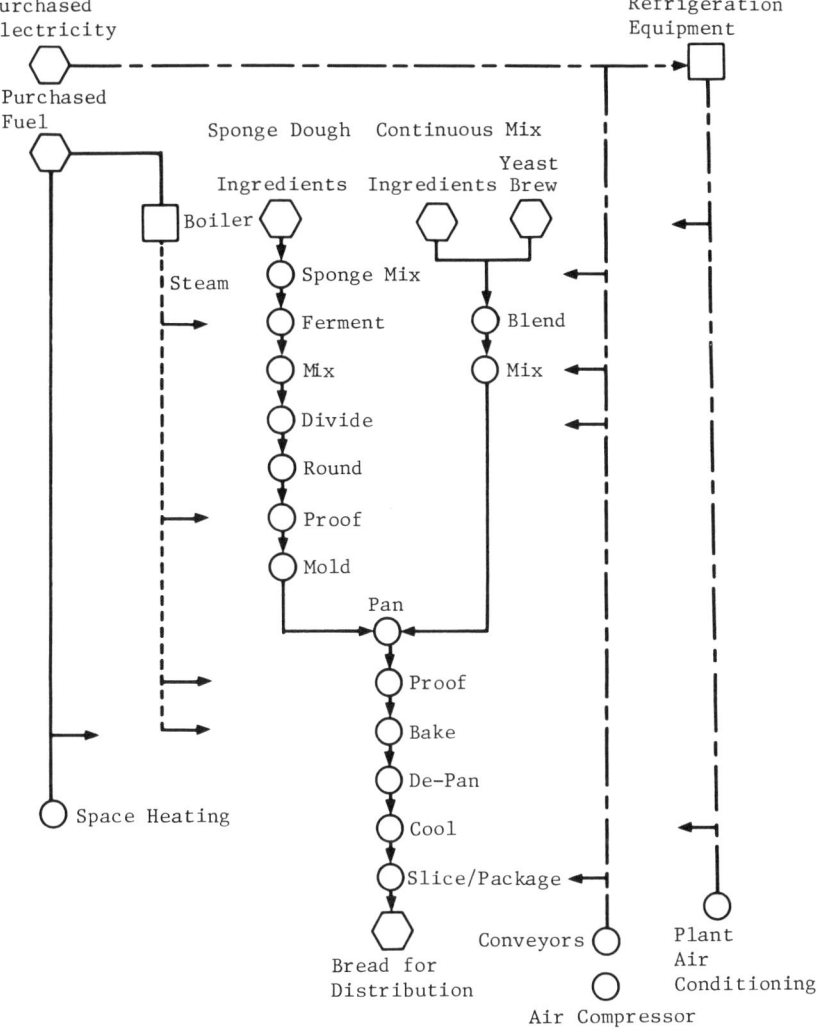

Figure 34. Material and energy flow, bread plant [DPRA 1976].

2. stoppage log report: list of stoppages for oven and wrapping machine and reasons for stoppage;
3. oven use report: available for the previous 24 hr (generated by request or automatically);
4. wrapping machine report: same as for oven use report.

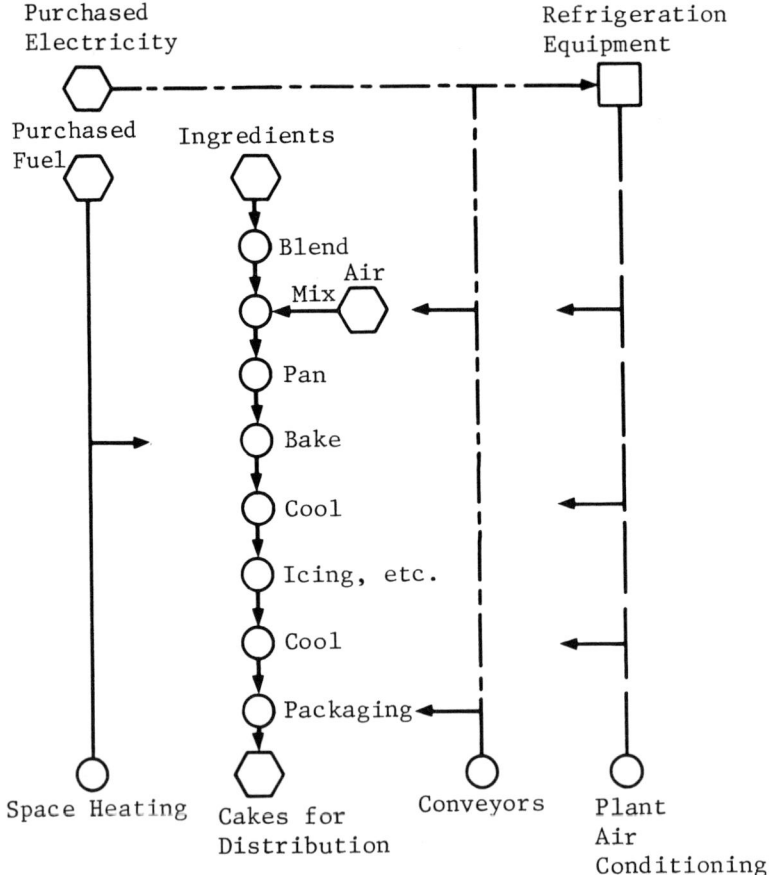

Food Industry

Figure 35. Schematic of APV feedback and highway system [DPRA 1976].

The turn-of-scale application is designed to measure and report weights of biscuit packets. Weights were previously recorded by operators. This system weighs with greater accuracy and automatically allows for crumb buildup. Data are continuously displayed at the oven-man's position. The production count application provides an accurate correlation of the volume of packets leaving the packing room and entering the warehouse. Factory floor message application permits messages between all areas of the factory. A log of all messages is available.

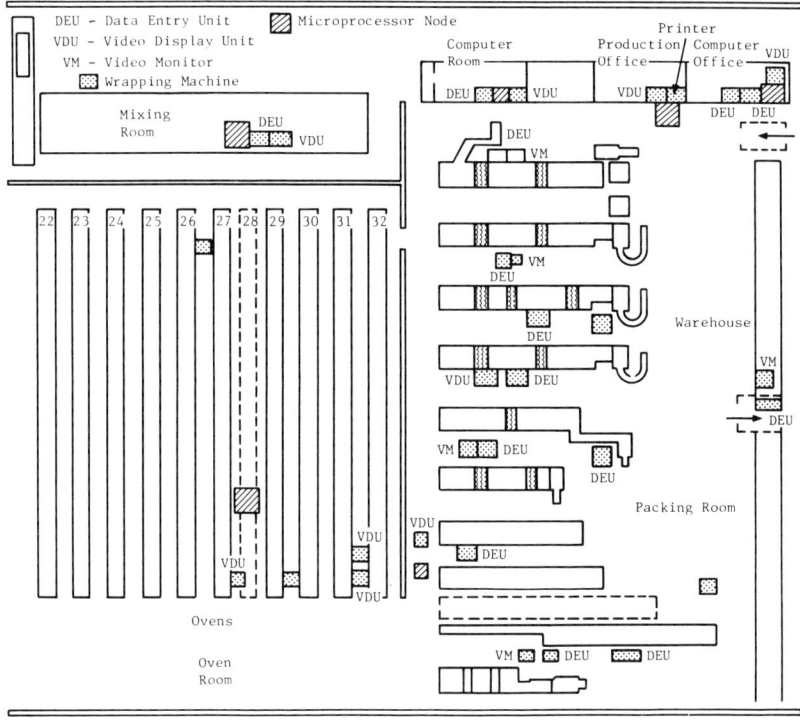

Figure 36. Floor layout of oven and packing rooms at United Biscuits' Harlesden (London) factory, showing ovens, and wrapping and packaging machines. Various types of terminals used in the production information system are distributed around the factory floor [Hope and Donlan 1978].

For production monitoring, the predicted cost and annual savings are $45,000 and $35,000, permitting a payback period of about 15 months. The cost of the check weighing application is $300,000; the annual savings of $250,000 allows for a payback period of 24 months.

Pastry Components

Henry and Henry, a company in Buffalo, New York, makes fruit and cream fillings, toppings, cake bases, icing bases and chocolate syrups. The former wet processing facility consisted of a fruit preparation department, and five separate multi- and single-cooking process kettle operations. Operations were inefficient and hampered by limited storage facilities, pumps, line and meter sizes. Instead of building a new plant, the facility was expanded, tank storage facilities were increased and new delivery lines were installed. A programmable process control system from Ladish Co. was installed.

All ingredients for 79 different products are controlled by programs that handle three liquid sugars, hot and cold water, liquid shortening and milk solutions. The system delivers ingredients into one of three 275-gal (1039-liter) kettles where the batches are cooked and blended. Cycles may vary from 10 to 25 min and from 140 to 240°F (60 to 115°C). The complete batch is pumped to a 600-gal (2268-liter) holding tank. This side tank allows the process to be continuous. From the holding tank, batches are sent to one of three filling lines (e.g., bulk, pail or can) through mode selection. The product may go through or bypass a homogenization process. The remaining processes (labeling, packaging and palletizing) are not involved with the control system.

With a minimum of added personnel, production efficiency and output increased: 20 batches can now be made on one line in an 8-hr shift, as compared to 16 with the manual system (a 25% increase). On another line, the number of batches has increased from 9 to 13 per shift (a 44% increase). The exact control of ingredients is improved by eliminating the human factor.

Downtime is costly in bakeries; one baker estimates a loss of approximately $100/hr. The previous electromechanical relay (EMR) system at Del Campo Bakery in Wilmington, Delaware, experienced frequent failures. The contact picked up a coating that interfered with proper operation. An EMR system that can handle 1000 lb (454 kg) of dough requires 15–20 timers, and numerous photoeyes. Operation was extremely labor-intensive. Dough was proofed (brought to a standard lightness) in a box or humidification room, wheeled in on racks holding 20 pans, and hand-loaded.

An automated system was installed with the conveyor and proofer having programmable control with solid state devices and no contact points. The operation is completely automatic. Continuous dough is placed on plywood "peel boards" taken by conveyor to the proofer, and transferred to stainless steel trays, which travel through the proofer on roller chains.

Reliability is improved; the conveyor and proofer have been down only once per year, compared with approximately a dozen failures with EMR systems. The automated system is simpler to operate, has fewer maintenance problems, and the controller saves space. All manual pushing and pulling of racks, loading, and unloading are eliminated.

Bottled and Canned Beverages

The beverage industry is composed of two major segments: alcoholic beverages and bottled and canned soft drinks. The alcoholic beverage industry is further subdivided into three major sectors: malt beverages (SIC 2082), wines and brandy (SIC 2084), and distilled liquors (SIC 2085). Manufacturers' shipments of alcoholic beverages and soft drinks were estimated to reach $14.5 and $14.4 billion, respectively, in 1980.

The malt beverages subsector is the most energy-consumptive, as indicated in Table XXVI. Table XXVII lists the percentage of total energy used by the major process steps.

The concepts of automation have been used at all process steps in the beverage industry. Major advances in the automation of the brewing processes and the monitoring of the product at each step were made between 1967 and 1977. The optimum degree of automation within a given facility is determined by: the type of equipment to be controlled, the processing method and the availability of capital. The benefits include increased productivity, improved plant efficiency and safety. Automation has reached the stage such that, if a process variable can be measured accurately, it can be controlled accurately. Sensing devices are available to determine temperature, flow, pressure, level and solids weight. In addition, a large variety of instruments are available for carrying out almost every analysis, including carbonation, dissolved oxygen, specific gravity and turbidity. For example, at least two methods for automatic on-stream beer carbonation analysis have been developed. One method employs a semipermeable membrane [Gamacke 1968] that permits the diffusion of carbon dioxide out of the stream. A second is an automated sequencing of the Zahm-Hartung carbonation analyzer [Rolner 1970]. Depending on the manufacturer, sensing and readout are pneumatic, electrical or a combination of the two.

An example of complete automation by computer control of all routing operations for fermentation and maturation in a modern brewry through fixed piping using hygienic control valves is described by Harris and Irvine [1978]. The first such system was completed in 1973. The control system employs a central processor unit programmed in "Paracode," a process language developed by the APV Company using what they have termed the "Highway" interface system (Figure 37). In this system, each plant item is assigned a unique number. A valve, for example, receives on-and-off drive signals to operate it, and feed backs to the control system open-or-closed position signals from microswitches. Full automation of in-place cleaning and routing compared with automanual

Table XXVI.   Beverages: Energy Consumption Patterns[a]

SIC Code and Year	Total (10^12 Btu)	Distillate Oil 10^12 Btu	%	Residual Oil 10^12 Btu	%	Coal 10^12 Btu	%	Coke 10^12 Btu	%	Natural Gas 10^12 Btu	%	Other Fuels 10^12 Btu	%	Purchased Fuels 10^12 Btu	%	Electricity 10^12 Btu	%
2082																	
1971	47.9	3.6	7.5	7.4	15.4	4.4	9.1			26.4	55.1	0.9	1.9	42.6	89.0	5.3	11.0
1972	50.1	4.0	8.0	7.5	15.0	4.0	8.0			28.1	56.0	1.0	2.0	44.6	89.0	5.5	11.0
1974	49.5	5.9	12.0	5.0	10.1	D^b				27.1	54.8	D		43.5	87.9	6.0	12.1
1975	46.7	5.7	12.2	6.4	13.6	D				24.3	51.9	D		40.3	86.1	6.4	13.9
1976	46.8	6.9	14.7	6.6	14.1	D				22.6	48.3	D		39.8	85.0	7.0	15.0
2083																	
1971	9.3		1.0		0.9		1.7			8.0	85.9		2.2	8.5	91.7	0.8	8.3
1972	9.4		1.0		1.0		2.0			8.0	85		2.0	8.6	91.0	0.8	9.0
1974	9.2	0.2	2.2	D		D		D		7.6	42.4	D		8.2	90.3	0.9	9.7
1975	9.6	0.3	3.4	D		D		D		7.6	79.3	D		8.5	89.3	1.1	10.7
1976	9.4	0.4	4.4	D		D		D		7.1	75.2	D		8.3	88.3	1.1	11.7
2084																	
1971	4.0	0.2	4.7		1.5					3.1	77.2		2.0	3.4	85.4	0.6	14.6
1972	3.5	0.2	5.0		1.0					2.6	75.0		2.0	2.9	83.0	0.6	17.0
1974	4.1	0.2	5.1	D						1.9	47.5	D		3.2	78.1	0.9	21.9
1975	4.1	0.1	3.3	D						2.4	57.5	D		3.1	75.0	1.0	25.0
1976	4.2	0.1	3.2	D						2.4	56.1	D		3.3	78.6	0.9	21.4
2085																	
1971	21.3	1.1	4.1	1.5	7.2	13.7	64.0			4.4	20.6	D		20.5	95.9	0.8	4.1
1972	21.4	1.1	5.0	1.5	7.0	13.5	63.0			4.5	21.0	D		20.5	96.0	0.9	4.0
1974	17.1	1.5	8.5							7.3	42.5			16.1	94.4	1.0	3.6

1975	13.6	1.6	11.8	D	3.6	9.2	5.5	D	6.7	12.6	92.5	1.0	7.5
1976	15.5	2.0	12.8		3.0	10.0	5.7	D	6.0	14.0	90.3	1.0	9.7
2086													
1971	28.7	3.6	12.5	0.9	2.6		53.7	1.9		24.6	85.7	4.1	14.3
1972	30.2	3.6	12.0	D	3.0		55.0	1.8		26.0	86.0	4.2	14.0
1974	20.1	1.8	9.0	D	D		36.2	D		16.4	81.3	3.7	18.7
1975	20.5	1.4	6.8	D	D		36.5	D		15.7	76.7	4.8	23.3
1976	20.9	1.0	4.7	D	D		37.7	D		16.3	78.0	4.6	22.0
2087													
1971	7.4	1.3	17.4	2.1	28.0		38.8		8.5	6.8	92.7	0.6	7.3
1972	7.9	1.0	13.0	2.1	27.0		44.0		6.0	7.1	90.0	0.8	10.0
1974	5.1		4.1	1.3	25.6		54.0		3.4	4.2	87.3	0.9	12.7
1975	3.8	0.1	3.9	0.9	25.3		40.9	D		3.1	81.8	0.7	18.2
1976	4.4	0.3	5.7	1.1	25.8		34.9	D		3.6	81.8	0.8	18.2

[a] Data for 1974–1976: DOC [1976–1978]. Data for 1971 and 1972: DPRA [1976].
[b] D = withheld to avoid disclosing information for individual companies.

Table XXVII.   Energy Distribution, Percentage Use
by Process [DPRA 1976]

Use Category	SIC Code					
	2082	2083	2084	2085	2086	2087
Direct Fuel Use						
Drying	4.2		4.0			
Space heat	1.0	4.5	2.0	7.7	7.5	4.7
Process heat	1.0					
Transport	1.8		2.0		4.5	
Kiln		85.5				
Cooping				0.3		
Figal can washing					3.0	
Evaporation						83.7
Preparation						4.7
Boiler Losses	18.6		19.3	20.3	16.8	
Steam						
Process	42.9			39.8	24.8	
Plant generation	4.9					
Other losses	9.7		3.4	9.4	4.7	
Hot pressing			1.5			
Distillation			60.8			
Bottle washing					20.8	
Electricity						
Mechanical power	4.5	9.5	7.4	11.3	3.2	6.7
Refrigeration	3.0		5.8		5.2	
Lights	3.2	0.5	0.8	2.7	5.2	0.3
Transport	0.3					
Battery charging					2.9	
Space heat	4.9			8.6		
Air-conditioning					1.4	

increases capital costs in single tank operations by less than 1% and in two tank operations by less than 5%. The Kronenbourg facility in Obernai, France, is also an outstanding example of an automated brewery. The annual output is 600 million liters with a staff of only 1050. The brewery's performance figures include: 0.03 kg of fuel, 0.08 kWh and 7 liters of water per liter of beer. All of these improvements are attributed to plant automation.

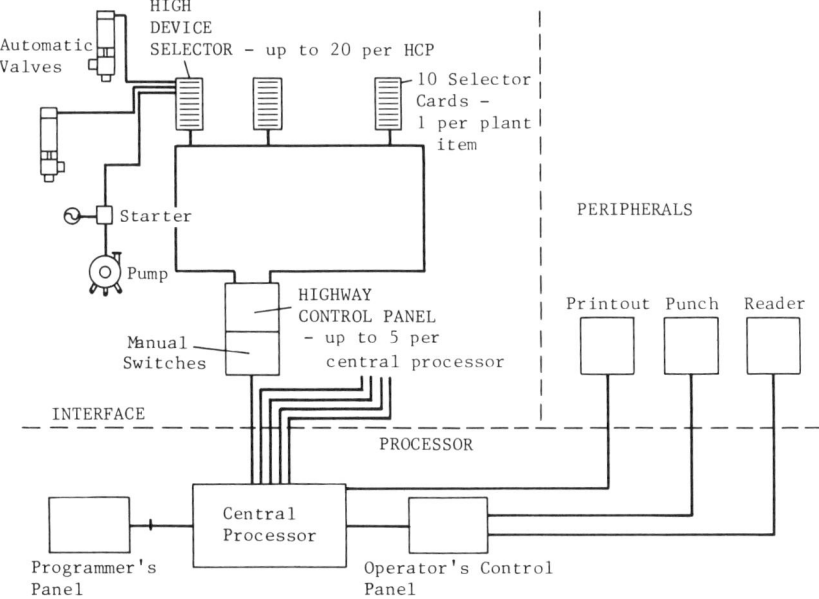

Figure 37.   Material and energy flow, cake plant [Harris and Irvine 1978].

PULP AND PAPER INDUSTRY

Industry Characteristics

The pulp and papermaking industry is important in the United States:

- It ranks fourth in the country as a consumer of fuel and power and first as a manufacturing consumer of fuel oil (2.30 quadrillion Btu, or 24.2×10^{17} J, in 1972).
- It ranks among the 10 largest industries in the United States with a 1979 output of $65 billion, 6000 plants in 49 states and 685,000 employees. The pulp mills industry alone had shipments valued in 1979 at $2.664 trillion and it employed 16,200 workers in 45 establishments.
- It is the third largest user of water.
- Of the total energy the industry consumed in 1976, 41.4% was derived from self-generated and waste fuels.
- It ranks ninth among U.S. industries in capital intensity.

Electronic computers are increasingly used for process control in the pulp and paper mills. More than 150 process control computers were used in the pulp and paper industry in 1973, compared to 11 in 1965. Computers improve control of papermaking machines, pulpmaking equipment and bleaching systems.

A process flowsheet for an integrated bleached draft pulp and paper mill is provided in Figure 38. Opportunities for computerized automation exist at various points during this process. The processing steps and automated systems will be discussed below, beginning when wood is processed into pulp.

Pulp Processing

The use of computers in the pulp industry has been reviewed by Lavigne [1977,1979]. The primary incentives for the use of process computer control systems are:

- manufacturing cost reduction;
- improved production efficiency;
- improved product quality; and
- safer process operating conditions.

Pulp production begins when round wood (logs) is debarked, then reduced to chips of approximately $1 \times 0.5 \times 0.25$ in. $(2.54 \times 1.27 \times 0.635$ cm), using a rotating flywheel faced with bars that act as cutting blades. The chips are screened, and the fines are sent to a boiler combustion unit as a fuel with the dominant draft process. The accepted chips are then sent to a pulping digester where they are cooked under pressure with a cooking liquor. The digester may operate continuously or process the chips in batches. The cooking liquor, or white liquor, consisting of an aqueous solution of sodium sulfide and sodium hydroxide, dissolves the lignin that binds the cellulose fibers together. When cooking is completed, the contents of the digester are forced into the blow tank. Here the major portion of the spent cooking liquor, which contains the dissolved lignin, is drained, and the pulp enters the initial stage of washing. From the blow tank the pulp passes through the knotter, where unreacted chunks of wood are removed. The pulp is then washed and, in some mills, bleached before being pressed and dried into the finished product. For integrated pulp and paper mills, the pulp drying step is eliminated. Finally, it is economically necessary to recover both the

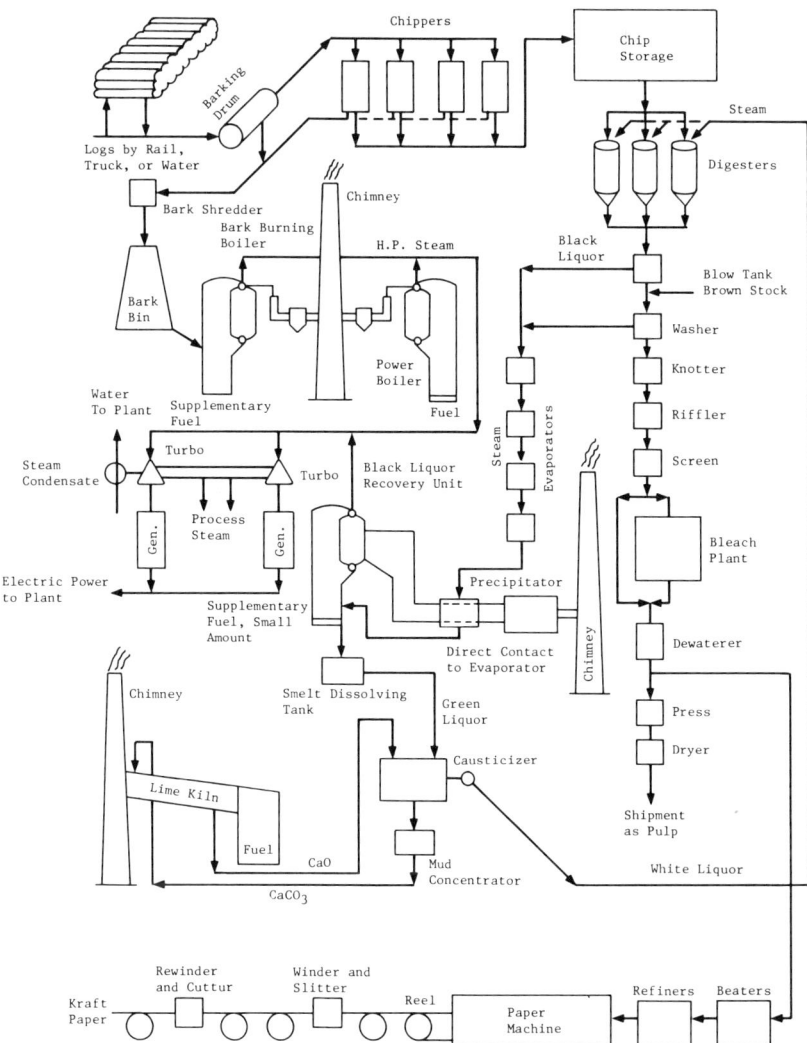

Figure 38. Flowsheet for kraft pulp and paper manufacturer.

inorganic cooking chemicals and the heat content of the spent black liquor, which is separated from the cooked pulp. The liquor is concentrated and then sprayed into the recovery furnace, where the organic content is burned. The green liquor is then conveyed to a causticizer where it is treated to be reused in subsequent cooks.

Pulp Mill Automation

Thompson [1978] made a comprehensive analysis of the potential savings associated with automating every step of a pulp mill. He found potential savings at all steps, with an improved digester yield, a decreased need for cooking chemicals, steam optimization, improved turpentine recovery, increased production, improved quality, more efficient bleach plant operation and steam saving. Table XXVIII provides an economic balance sheet between the investment and operating costs associated with automation and the annual savings derived. The balance is attractively positive.

Pulp Digester

The pulp digester represents an important point in the process where computerization can play a valuable role. In the manual operation of a batch pulp digester, the same amount of makeup liquor is customarily used, regardless of the chip moisture content. The wetter chips require more steam to heat to temperature than would be necessary if the makeup liquor were properly adjusted for chip moisture. Since it is not possible to monitor accurately and continuously the moisture content of the chips by manual methods, there are inconsistencies from batch to batch. The problem is accentuated in directly steamed batch digesters, since in these units some of the steam entering the digester will condense, changing the liquor-to-wood ratio.

The deficiency can be alleviated with an automated batch digester, shown schematically in Figure 39 [Powell 1979]. The charge control supervises the digester filling operations, checking that the correct amounts of chips, cooking liquor and makeup liquor are added to the digester. Cooking control brings the charge to the appropriate temperature and cooks it according to a prescribed time-temperature relationship (H factor). Cooking control permits a reduction in the typical manual operating practice of having safe margins of temperature and time.

The typical control functions performed by a batch digester computer control system include:

- chip charging;
- liquor charging;
- steam control using a factor correction;
- K (kappa) number control;
- relief control;
- circulation control;

Table XXVIII.   Savings from Pulp Mill Automation ($1000s): Pulp Mill Subsystem (Figures Estimated)
[Thompson 1978]

Description	Capitalized Costs					Annual Expenses		
	Wiring Instrum.	Start-up Manpower	Programming	Misc.	Total	System Rental	Continuing Manpower	Annual Savings
Pulp Mill System		4.5	20.0	10.0	34.5	30.0	12.0	
Wood Pref	13.6	1.5	8.0	1.0	24.1	17.2		
Digesters	77.1	6.0	45.0	5.0	133.1	13.2		418.8
Wash/Screen	35.3	3.0	10.0	3.0	51.3	8.4		4.8
Bleach Plant	58.7	8.0	35.0	4.0	105.7	12.0		228.0
Totals					348.7	70.8	12.0	651.6

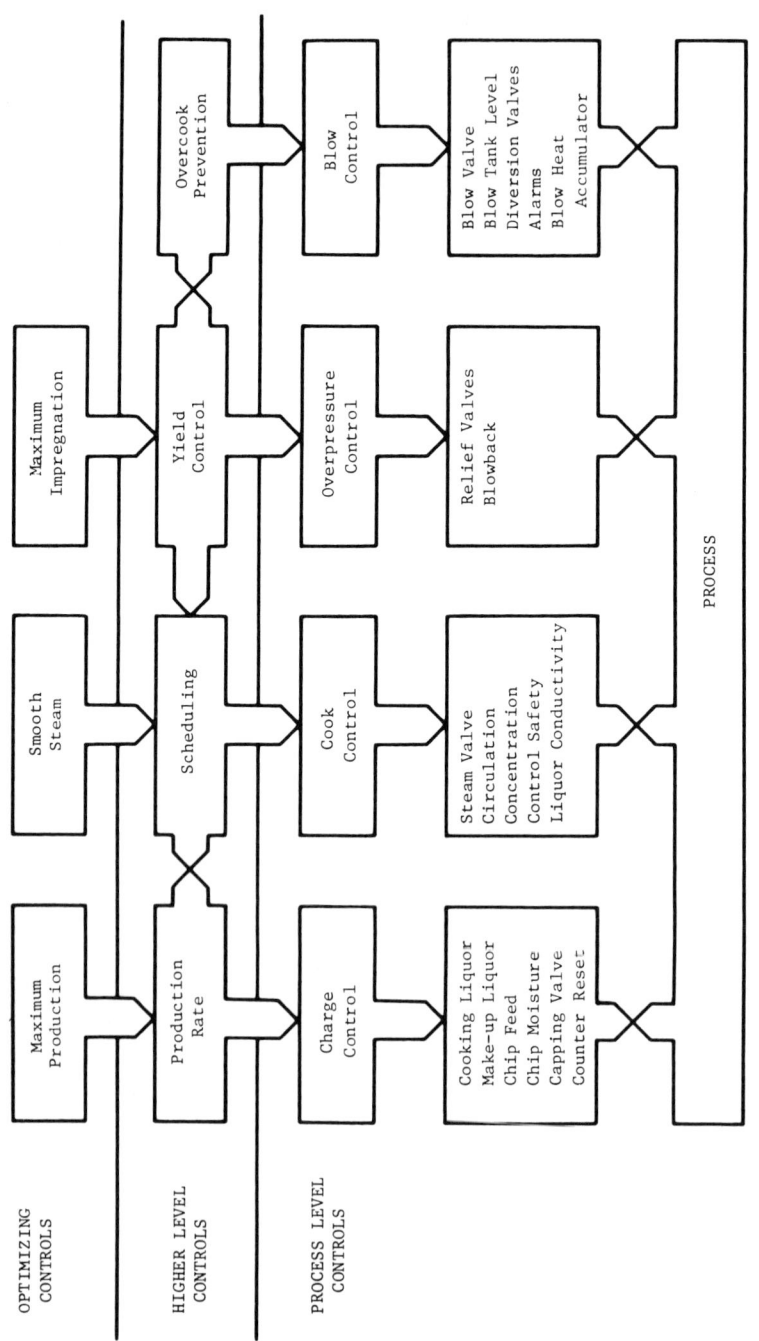

Figure 39.   Batch pulp digester control strategies [Powell 1979].

- blow control;
- automatic blow-back;
- production change coordination;
- blow heat recovery; and
- steam load leveling.

The automated system has the following advantages:

- reduced alkali-to-wood ratio;
- elimination of production limit resulting from recovery cycle;
- increased digester packing efficiency;
- bleach plant chemical savings; and
- production increase.

Figure 40 summarizes the steam savings through computerized process control, assuming:

- steam cost of $3–4/1000 lb ($6–9/1000 kg) of steam;
- steam use of 3500 lb/ton (1723 kg/metric ton) of pulp;
- a 500-ton/day (454-metric-ton/day) operation;
- a value of steam of $12/ton ($13/metric ton) of pulp; and
- an average saving of 13.9%.

We can translate these steam savings into a savings of almost $300,000:

$$\$12/\text{ton} \times 500 \text{ ton/day} \times 350 \text{ day/ton} \times 13.9\% = \$291{,}900$$

Another benefit of automation is the reduction on steam swings from the powerhouse associated with a smoother demand. Steam swings are reduced an average of 77%. Such reductions improve powerhouse efficiency, causing a 35% demand reduction in boiler steam.

The main objective of computerized process control in the digestion step is the production of a pulp of uniform quality. The MODO Pulp Mill in Husum, Sweden, has been automated toward that goal [Fadum 1980; *Tappi* 1979] as shown in Figure 41. The most important variables are cooking time and temperature, and the concentration of the cooking chemicals. These variables and others are controlled, allowing one to dispense with the use of predictive models (normally used) for the temperature in the digester. The system has had an availability of close to 100% since its installation.

It is well known that better control over quality, as measured by the kappa number, provides a more uniform pulp; a shift in the kappa

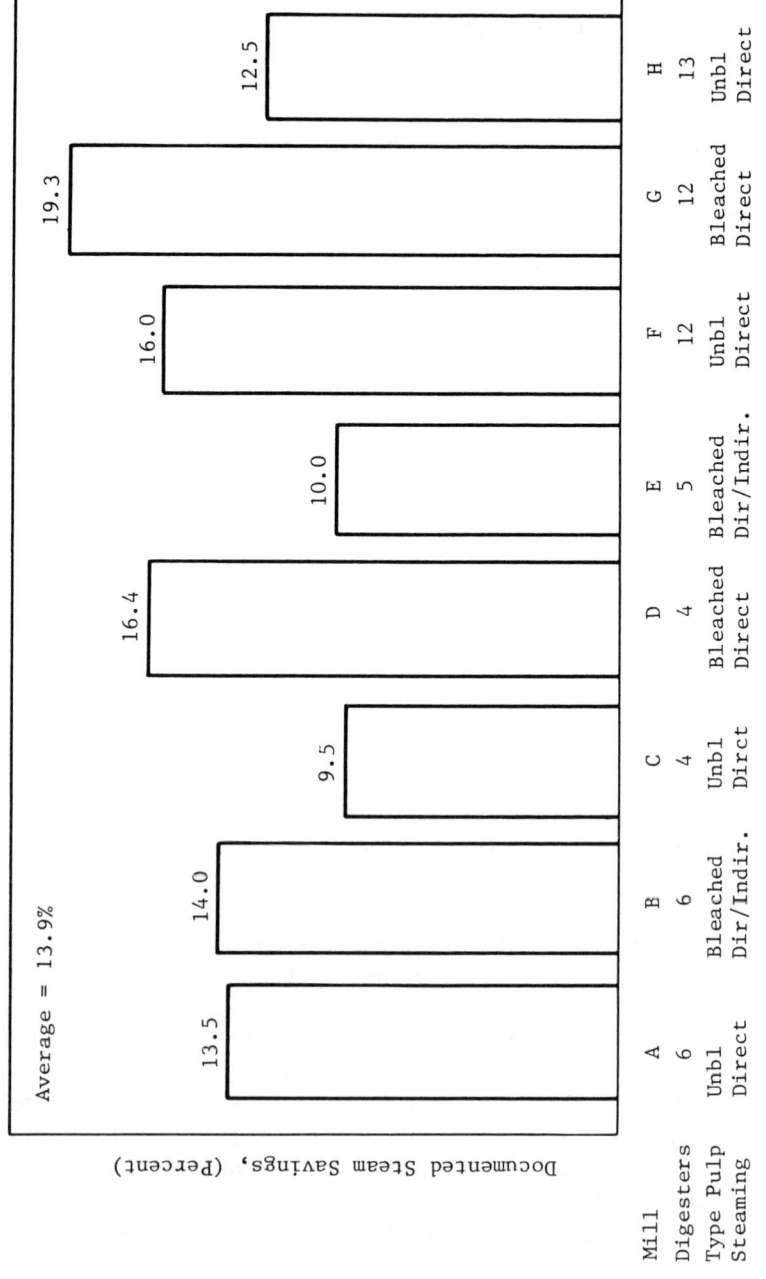

Figure 40.   Steam savings with automated batch digester [Powell 1979].

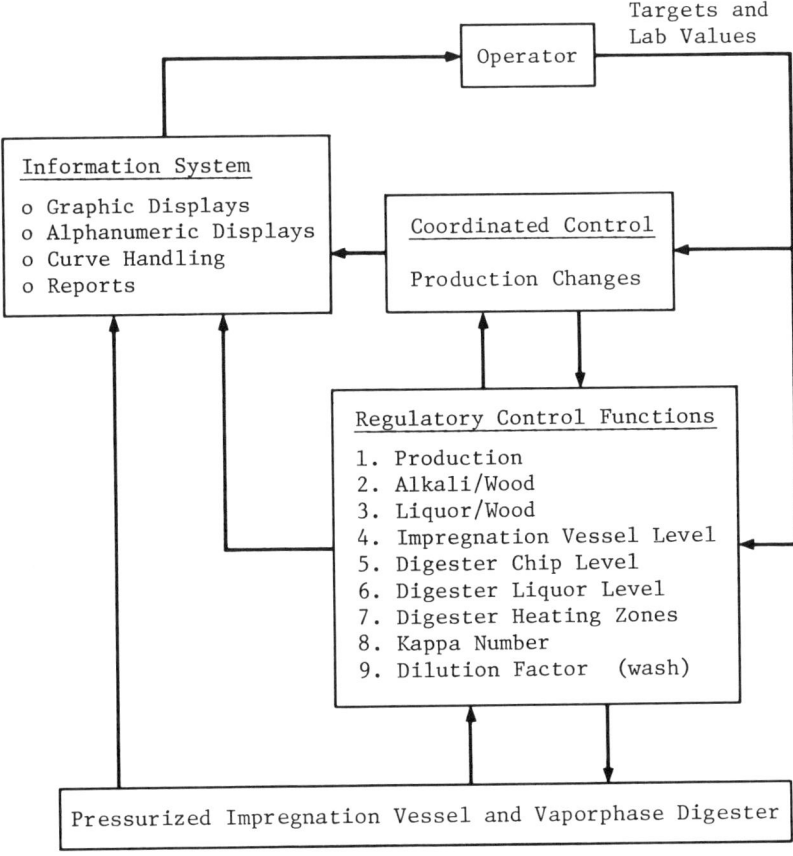

Figure 41.   Digester control program structure [Fadum 1980].

number can be used to achieve economic gain. The Swedish plant shows a weekly kappa number standard deviation of 1.3 on a base of 33 and as low as 0.89 on a 24-hr period.

In Varkaus, Finland, a thermochemical pulp mill has been automated [Metsavirta and Leppanen 1980]. The process control strategy is shown in Figure 42. Computer control provides process improvements and optimal energy consumption. The Varkaus system features controls on (1) production rate, specific energy consumption and power split in each refining line; (2) final pulp properties; and (3) reject rate. It can be shown that automation saves 1.6% of the energy used in the plant.

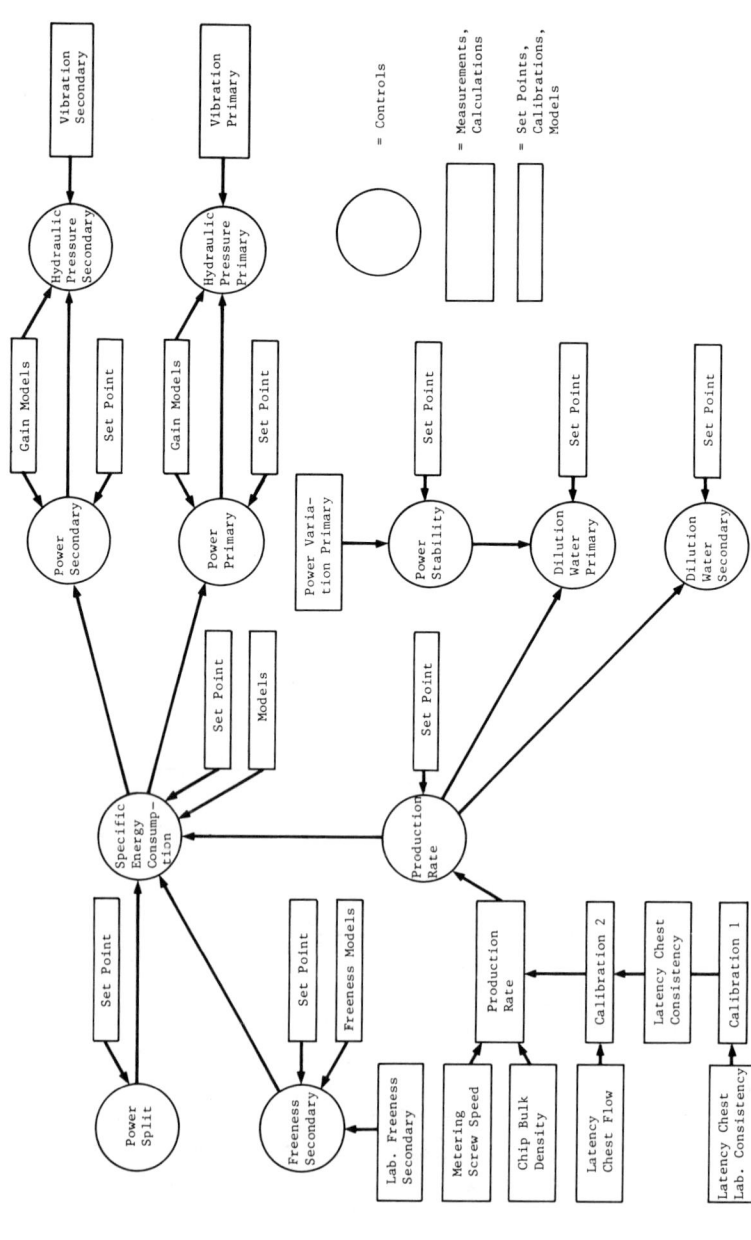

Figure 42.   Varkaus computerized process control strategy [Metsavirta and Leppanen 1980].

Pulp Bleaching

Pulp bleaching is another step where computerized automation can improve efficiency. It involves treating the wood pulp with chlorine gas to neutralize color agents that will affect the whiteness of the pulp. Subsequent treatment of the chlorinated pulp involves treatment with caustic, steam, bleach and rinsing operations. The amount of chemicals used depends largely on the amount of pulp to be processed and the amount of impurities. Traditionally these have been measured, but not with a high degree of accuracy, as that would require stopping the production line to do laboratory testing.

Figure 43 is a schematic diagram of an automated bleach plant. The automatic process control system monitors and controls all of the stages of pulp bleaching: chlorination, extraction, hypochlorination and chloride dioxide treatment. In addition, the system monitors the incoming

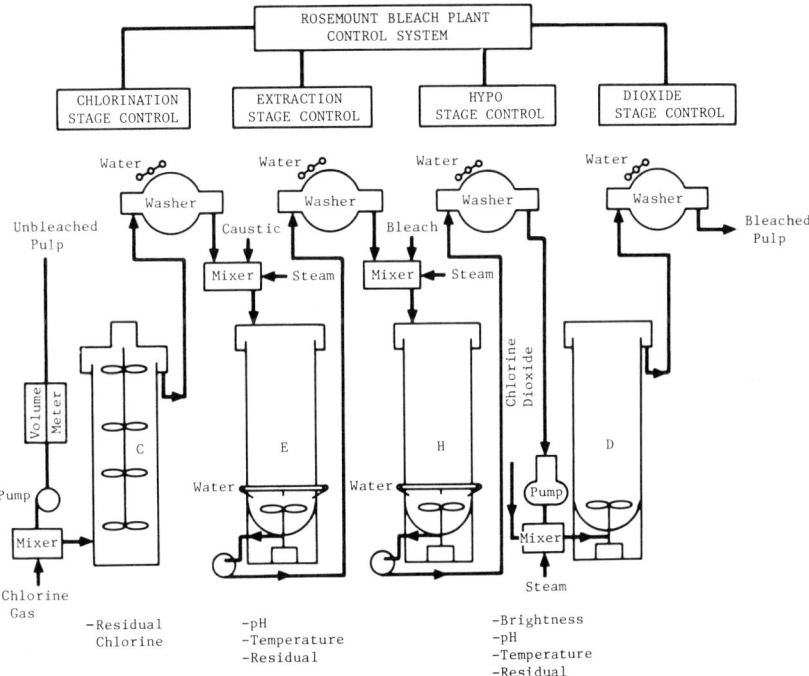

Figure 43.   Automated bleach plant. Benefits: lower chemical use, lower steam use, reduced brightness variations, improved pulp viscosity, increased production due to less shrinkage.

unbleached pulp to determine whether there are proper amounts of chemicals and steam. This offers the advantages of lower steam and chemical usage, improved pulp viscosity, less shrinkage and reduced brightness variation.

Pulp Drying

Pulp drying, the final step in the pulpmaking process, is accomplished with conveyor dryers. These often do not operate at full capacity because of limitations imposed by other process machinery, the production schedule or the running of a variety of products. Although drying rates for the material have been determined beforehand, variations in moisture content are assumed to be so small as to be negligible. Thus, the material to be dried is underdried or overdried, with the consequent loss of product quality or increase in energy use.

Computer control of the drying operation produces tighter control than would be possible with manual control. The computer monitors the moisture content of the dried material against a set specification. It monitors the drying rate so that the heat input to the dryer is closely matched with the heat requirements under the given moisture conditions. Figure 44 presents typical energy reduction and cost savings for a drying unit operating with and without automated process control [Zagorzycki 1979]. These controls result in lower operating costs per unit of material to be dried and decreased energy consumption by reducing unnecessary evaporation. Productivity is increased due to a decrease in the number of rejected units. There is an increase in product quality and uniformity.

Papermaking

After the pulp is made, it is processed into paper through four major mechanical steps: preparing the stock, forming the sheet, removing the water and finishing the sheet. (The same is true in the manufacture of paperboard.)

The fibers produced in the pulping process are mixed with water to form a thin mixture containing 1 part fiber to 200 parts water. This paper stock is further refined in a process known as beating. The pulp is passed between two sets of bars forming two rubbing surfaces. In this process, the fibers are shortened, the ends are shredded and the outer walls are ruptured.

Mineral fillers are often added to the stock; these fillers increase the paper's opacity, improve printing quality and increase brightness. The

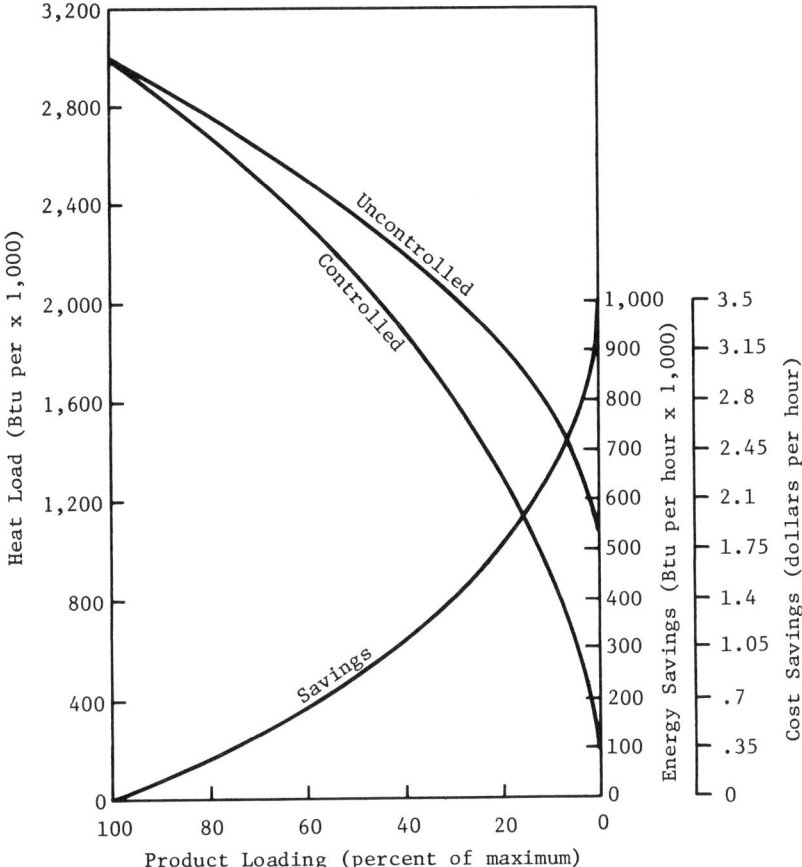

Figure 44.   Heat loads with and without humidity control and energy/cost savings [Zagorzycki 1979].

more important fillers are clay, calcium carbonate and titanium dioxide. Finally, to retard the flow of ink in the paper, a "sizing" material is added to the stock.

On completion of the stock preparation step, the stock is fed into a papermaking machine for forming the sheet. During the forming step, water is removed from the prepared stock in a multistep process that uses gravity, capillary action, and vacuum and mechanical presses. Still more water is removed in the drying section, which is made up of a series of large-diameter, hollow rolls that are heated with steam. The paper is passed over these rotating drums, supported on a system of felts designed to ensure continuous contact between the sheet and cylinder surface and to carry away from the sheet part of the evaporated moisture. After leaving the dryer section, the sheet is only 6–7% water by weight.

The dried sheet enters the finishing section, where it is passed through the calendar section. Here it is compressed to give it a smooth finish. After the finishing operation is completed, the paper is wound onto a reel, where it can be trimmed to the desired width. It should be noted that papermaking is a continuous process.

Like pulpmaking, papermaking can be accomplished more efficiently with automation. At one plant studied by the Bureau of Labor Statistics [DOL 1975], the addition of a process control computer to a paper-making machine reduced grade changeover time by 20%, increased machine speed by 15% and improved machine efficiency by 2% for an overall gain in production of 19%.

The majority of computers in the paper mills are concerned with stock preparation and the optimal functioning of the paper machines. Computer control in stock preparation is used during the refining and blending operations to minimize variations in the stock. In stock blending, computers are on-line to define appropriate proportions of dry weights of pulps, chemicals and additives to achieve mill demand requirements.

The paper machine step offers the greatest influence on the properties of final paper product; hence, the first computer applications were found at this step. The computer can handle grade changes by referring to stored standards and modifying running conditions for the desired grade. Computers are also used to monitor the various machine dryer sections and to advise operators on required changes in running condition. Computers have also been used in color control of the paper sheet. Finally, computers are available to control paper trimming machines and the speed of paper mill drives [Lockley 1979].

WATER DISTRIBUTION AND SEWAGE TREATMENT*

Automated Water Distribution Systems

Many U.S. cities have automated water distribution systems, the Ohio River system [ORVSC 1970] being one of the earliest and most sophisticated. The application of automation methods to the water utility and

---

*Ouellette et al. [1975] extensively reviewed the use of computers in environmental monitoring and controls.

sewage treatment processes is characterized by the use of centralized control centers. Remote measurement transducers report data to a centralized facility where a computer monitors the process. Instructions are issued to activate electrically operated equipment under control of a human operator. Examples of automated projects were listed by a major manufacturer in the June 1980 issue of the *Brown-Boveri Review*. [Baensch et al. 1980; Flotho et al. 1980; Herold 1980; Steele 1980].

The city of Augsburg, Germany, installed a sewage treatment plant with a capacity for a city of 960,000 that is controlled and monitored by a process computer [Baensch et al. 1980]. The sewage treatment process involves aerobic treatment of the sewage and anaerobic digestion of the resulting sludge. The overall control and monitoring system is divided into three levels:

1. process measurement of level, flowrate, pH, etc.;
2. group control, which activates equipment related to the process measurements; and
3. a central control room where overall process control is performed.

The central control room communicates with the system through a keyboard and CRT display. In the event of computer failure, the commands are given from a display board. The transducers required to measure the process parameters are an important and difficult part of the system, because of the problems of reliable transducer operation in the wastewater. Some solutions to the difficult instrumentation problems include:

*   an acoustic echo sounder for measuring water level;
*   capacitive transducers for measuring sludge levels;
*   ultrasonic measurement of sludge density;
*   light absorption measurement of dried matter in the activated sludge tanks; and
*   infrared absorption analysis of the digester gas.

The process computer collects 300 analog and 1500 binary signals on operating states. On-line, closed-loop process control, and, ultimately, process optimization with energy conservation are major future goals.

Municipal systems for pumping and distributing water often have a complex network of pumps, pipelines, storage reservoirs and water towers. The problems of keeping the required pressures, meeting demand and optimizing performance are solved using a process control computer. Pressure losses increase as the square of the flowrate, so energy consumption is reduced if high pumping rates are avoided. Because the efficiency of pumps varies with the operating pressures and flowrates, energy savings of 20% can be obtained by optimum selection of the pumps in use and maintaining low flowrates.

A computer system to control the water supply must:

- acquire and analyze current data from remote locations;
- predict the performance of the system based on historical data;
- simulate the water supply by a mathematical model; and
- perform step optimization to find the best control strategy.

Systems of this type have been implemented successfully in the United Kingdom with results that definitely justify the costs. Experience has shown that accurate and reliable data-gathering equipment is required for satisfactory operation [Luton 1978].

In a modernization of the water supply for the city of Stuttgart, Federal Republic of Germany, a centralized computer was installed that controls more than 200 wells that are part of the supply system [Flotho et al. 1980]. The control center consists of a small room with two color CRT displays for system display and one black-and-white display for data. Two typewriters print out permanent records and list disturbances that require prompt attention. Disturbances can be simulated for training purposes without interrupting the operations. The supply and demand logs are recorded and consumption is predicted for a complex system involving 56 pressure zones, and numerous intermediate reservoirs and pumping stations.

Automated Sewage and Treatment Systems

The automation of plant effluent treatment plants is not as common as the automation of water distribution systems. Two cases are typical. The first is a wastewater treatment facility at the Long Island Lighting Company handling intermittent process effluent flows averaging about 500,000 gal using programmable controllers at minimum cost [Baker and Rahoi 1981]. The second is a more sophisticated system installed at the Davyhulme sewage work treating sewage and industrial effluent from the Greater Manchester (England) Conurbation. The system uses a central computer collecting raw data by telemetry and providing:

- supervision of the plant utilities mimic, data and alarm displays;
- alarm, event and routine logging;
- plant sequence control; and
- plant modulating control [Windle and Chappell 1981].

INDUSTRIAL BOILERS

Background

Computers are increasingly used in energy management by energy-intensive industries requiring large amounts of thermal, electrical and mechanical energy. Areas in which the impact of computers is being felt include:

- steam and electricity use accounting;
- economic boiler dispatching;
- steam and electric load balancing; and
- boiler efficiency and performance monitoring and control.

Energy accounting is practiced, for example, in large pulp and paper mills. This accounting helps managers to reduce energy consumption.

A computer program for multiple-boiler economic energy dispatching can help in minimizing fuel cost. This is especially true in the pulp and paper industry, which uses a considerable amount of sawdust and bark, as well as recoverable black and red liquors [Lavigne 1979].

When electricity and steam demands are too large by comparison with the available in-plant production and purchase energy, load balancing between steam and electricity can be performed by computers assisting in the optimal energy redistribution.

Boiler efficiency and performance monitoring is accomplished using true process computers to optimize combustion efficiency [Jandorf 1981; Williams and Rogers 1981].

For a given fuel, one of the most important factors in determining efficiency of boiler combustion is the amount of oxygen available. Insufficient air and poor mixing result in combustible material being lost through the stack. Figure 45 shows the relationship between heat loss and the percentages of excess air and oxygen.

The typical solution to this problem is to have a device that can alter the fuel-to-air ratio [Luton 1976]. In small boilers, little attention has been paid to this problem. Evidence indicates that a 5% improvement in efficiency can readily be achieved.

There are several ways of improving the operating efficiency of industrial boilers. The oldest is to change the air-to-fuel ratio and then run outside to see if the smoke turned "white." This, of course, is restricted to coal and oil; with gas no visible effect is observed. Another way of improving the efficiency of boilers is to have the operator change the settings by checking the operating performance of the boiler as reported in time charts. This requires the operator to periodically check the perform-

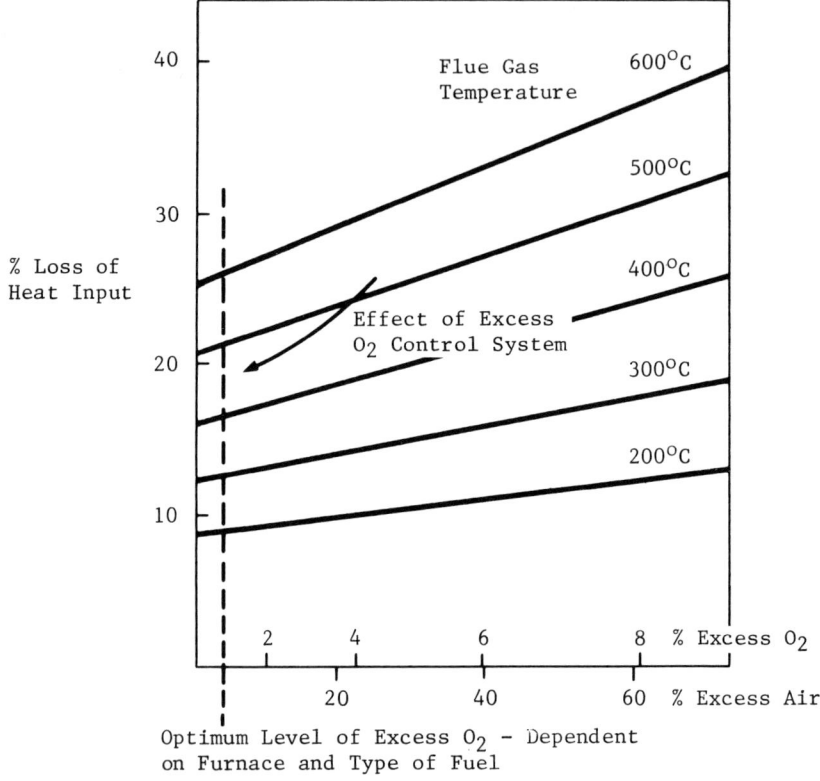

Figure 45.   Steam boiler heat loss [Luton 1978].

ance of the boiler. To assure satisfactory operation, the boiler is usually operated with an excess of air, which results in heat losses since the excess air draws heat from the combustion gases.

Automation

Figure 46 [Spanbauer 1980], shows a schematic diagram of an automatic boiler control. The process control system monitors the carbon monoxide concentration in the flue gas, maintaining it at a prescribed level adequate to ensure complete combustion, but not to heat excessive air. The CO concentration data are used to set the air-to-fuel ratio that optimizes the efficiency of the boiler. In addition (not shown in the figure), the control system allocates changes in steam demand to the

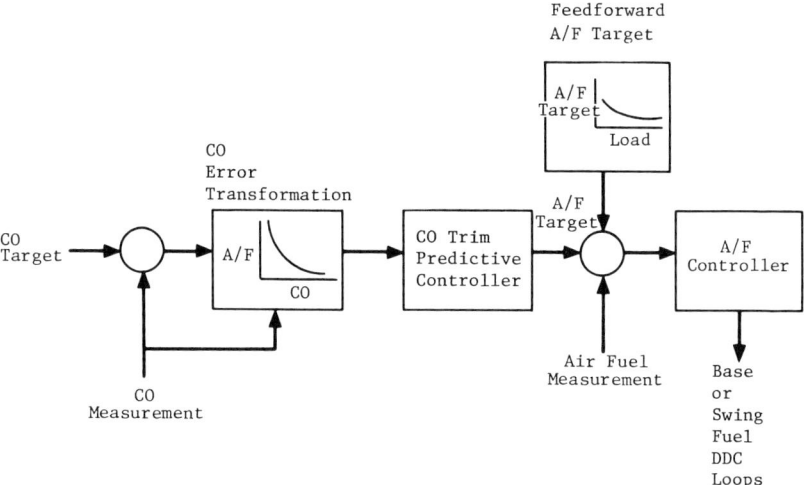

Figure 46. Excess air optimization strategy. A/F = air/fuel ratio; DDC = direct digital control [Spanbauer 1980].

most cost-effective boiler and optimizes the soot-blowing strategy based on actual fouling of the heat transfer surfaces. The computerized system has the following advantages:

- Computerized CO control strategies can cut stack heat losses by as much as 5%.
- Fuel consumption is reduced because less fuel is wasted.
- A computerized energy management system can result in a 6–8% savings in purchased energy.

Another application is the use of supervisory computers to improve energy consumption associated with cold cycles [Lucet et al. 1980]. Cold cycles that remove heat at different temperatures with one or more compressors require a large amount of energy. Algorithms are available to generate cold cycle setpoints. Application in a propylene plant reduced energy consumption of the compressors by 20%.

Heat exchanger networks are common features of most petrochemical and chemical processing plants [Challand et al. 1981]. The problem of synthesizing optimum heat-exchanger networks dates back to the 1960s. The purpose of heat-exchanger integration or synthesis is to find the structure or layout that will minimize the necessary incremental investment and operating costs for the system. Basically, a system is composed of the major processing unit, a heat-exchange network, and an auxiliary heat supply or removal system.

Heat integration is rapidly becoming a popular means of conserving energy. Feed preheating by highly integrated heat-exchanger networks is becoming more common. Even a small improvement in a high-temperature process, however, can mean a substantial savings in energy. A benefit analysis of the integrated system is shown in Table XXIX.

OTHER INDUSTRIES

Process control computers are used in a variety of industries in addition to those discussed in detail in this chapter.

In the textile industry, computerized finishing integrates dyeing and finishing techniques to reduce unit labor costs and improve quality [Zeiserl 1973].

In the footwear industry, computer tape stitching and numerically controlled sewing systems permit rapid style change and greatly reduce the unit labor of skilled sewing operators [GAO 1980].

In the plastics industry, computers control systems, monitor or record extruder operating variables, and coordinate completely automatic extrusion process lines [Smoluk 1978].

In the hydraulic cement industry, central process control wiring of major production processes—especially raw and finished grinding, kiln and material storage—have led since 1974 to substantial manpower savings [DOL 1975].

Table XXIX.   Benefits of Process Synthesis
in Heat-Exchanger Integration

	Original System	Synthesized System
Heat-Transfer Area (ft^2)	114,000	163,100
Recovered Heat (million Btu/hr)	298	342
Furnace Duty (million Btu/hr)	240	196
Reduced Duty (million Btu/hr)		44
Reduction (%)		18.3
Fuel Savings[a] ($ million)		1.22
Investment ($ million)		1.85
Simple Payback (yr)		1.5

[a] For a cost of $3.30/10^6 Btu.

In the lumber and wood product industry, automated sawmills with electrosensors and on-line computers analyze every log and determine the best way to cut the log for maximum product yield and minimum waste [*Control Engineering* 1981].

In the tire and tubes industry, the use of computers for process control is limited largely to optimization of compounding formulas (composition of ingredients necessary for rubber fabrication to attain desired quality of rubber with a minimal loss) [DOL 1974a,b], and for calendaring (applying rubber to fabric cords).

In the aluminum industry, computer automation is used for a cold-rolled product mill and for continuous heat treating and annealing line.

Computers are used in the paint industry for such traditional functions as inventory and formula calculations. The most important use of computers is in color matching. Computers are used only by larger paint firms [Herman 1974].

In the soap and detergent industry, computers automated the instruments that monitor and control the process. Computerized jobs include inventory control, flow and measurements of raw materials, formula calculations and mixing operations [Arrowsmith 1978; Wilder 1978].

CHAPTER 3

CONCLUSIONS AND TRENDS

The following conclusions have been derived from analysis of the information collected during this study.

AUTOMATION

Automation will continue to be influenced by the general trends in the computer industry of smaller, less expensive systems, greater ease of interaction between computer and operator, and reduced computer-system caused downtime. Installation and software costs are larger than the cost of the computer itself. The availability of trained personnel who can program and operate computer systems is the largest factor determining the use of automation. Manufacturers of automation equipment will continue to develop systems that are easier to operate with factory-labor skill levels by using more interactive languages and improved displays and controls. Numerically controlled machine tools are widely available but constitute only a small percentage of the tools sold. This trend will continue because numerical control is expensive and most manufacturing organizations do not have an operation at which the increased production can support the costs. Highly automated production technologies cut across the traditional lines of management and organization of a factory. Traditional business organizations may have difficulty accepting the disruption of established patterns that accompany a high level of automation. Robot technology is widely used in the auto industry and is beginning to be used in aerospace. The auto industry is expected to retain half of the market for robots and begin using robots in automated pick-and-place assembly stations.

The completely automated factory is 10 years from receiving widespread acceptance in the United States. In such a factory, a hierarchical system of networked computers and microprocessors is used in the control and delivery of raw stock and parts in process, forming processes for metal (and other materials), production planning scheduling, mechanical assembly, inspection and testing, maintenance and repair, and accounting operations. Before this concept will be accepted by the manufacturing industry, several demonstration plants must be designed, constructed, and operated by technically advanced industries such as automobiles and aerospace. The results must demonstrate more economical operation than is possible with the traditional methods, and provide satisfactory resolution of organizational problems.

For application to large processes, smaller computers are being integrated into distributed data processing networks. Microprocessors are being incorporated into smaller units, down to the level of individual sensors. More widespread applications to smaller processes are predicted as: (1) costs and sizes of computer systems decrease, (2) interaction with the operator becomes easier, and (3) energy and materials costs that can be controlled by automation increase. Pneumatic process control devices, despite the advent of electronic devices, still hold the biggest share of the market. This, however, is not expected to continue past 1990 as electronic controls become more sophisticated and less expensive. The integration of computer-aided design (CAD) and manufacturing (CAM) into the automatic factory is a difficult, complex problem. It involves several production technologies, complexities in factory organization; labor, engineering and management relations; the introduction of new equipment and procedures unfamiliar to existing personnel; and high capital investment in new, untried manufacturing methods. Since the economic advantages of the automated system are not obvious for many applications, management will be reluctant to accept it.

The construction of highly automated factories and plants is influenced by government policy and taxation of new capital expenses. To use automated techniques effectively, an industry often must construct a new plant. Replacement of obsolete facilities with new plants that have more efficient equipment as well as automation does not presently occur in the U.S. steel industry because of tax disincentives. The development of standards on a national and international basis will proceed in government agencies, trade associations and professional societies of industries using automation. Widespread use of automation will be facilitated as acceptable standards are adopted for design, machine control and the interfacing of computer data.

In monitoring devices, the trend is toward faster and more accurate sensors. In particular, the industry trend is toward the use of electronic devices for temperature, and pressure, fluid flow and emission measurements.

INDUSTRIAL APPLICATIONS

Motor Vehicles

The auto industry is currently the largest investor in CAD; integration of design with production will increase use of computerized methods for parts production. The large investment made by the auto industry will determine trends in automation that will spread to other parts manufacturing industries.

Aerospace

The aerospace and automobile industries make extensive use of CAD; integration of CAD with CAM is beginning. Smaller factories will expand the use of CAD as systems of computers and software become widely available. The aircraft industry will develop innovative automated construction methods because assembly represents half the cost of an aircraft.

Electronic Components

Automated methods are in widespread use for the design, layout and assembly of components in the electronics industry. The method of assembly reduces costs by more than 50% and errors to an acceptable level. The electronics industry uses highly automated methods for the design and assembly of equipment, some of which are unique to the industry. Two large electronics companies are expected to market this fabrication technology commercially.

Metals

Computers and automation are used in all primary metal industries. The steel industry is the best example of extensive use of process control computers, with applications in coke-making, sintering, blast furnace operation, steelmaking, reheating, rolling mills, heat treating and finishing.

Chemicals

Automated processes have been reported for ethylene, ammonia, polyvinyl chloride and lube oil plants. Refineries are highly automated, and distillation units also benefit from process control. The benefits include cost savings, energy reduction, flexibility and product quality.

Glass

Direct digital control of a glass melting process was demonstrated as early as 1964. Some of the benefits of automation include reduced energy consumption, longer furnace life and improved operating control.

Pharmaceutical Products

Computers are used on-line to control chemical processes, pharmaceutical operations, inventories, analytical instruments and physiological tests. The benefits include cost-effective sequencing of the batch, improved overall plant efficiency, reduced plant energy consumption, and improved efficiency of chemical, pharmaceutical and fermentation operations.

Foods and Beverages

The industry makes extensive use of computers to lower labor costs and use space efficiently while improving the uniform quality of products and yield. Important segments of the industry that use computers include the dairy and baking sectors. Computers are used in dairies to monitor and control milk flow and determine the butterfat content of different products. In addition, improved filling machines are linked with casing and stacking equipment. In the baking sector, advanced technology has long been a part of cookie and cracker production. It is used to a limited extent in baking bread and cakes. The bread sector may adopt automation to a greater extent if rising labor costs make it cost-effective to do so.

Pulp and Paper

In the pulping process, computers are used to control batch pulp digesters with increased efficiency, chemical savings, production increase, steam savings and improved pulp uniformity and quality. Computer control of drying operations allows tighter operations than manual control with associated energy use reduction. Paper mills also have been automated extensively for grade selection, color control and trimming operation, with substantial savings in material and improved quality.

Water Distribution and Sewage Treatment

The municipal network for pumping and distribution of water is under computer control in many cities. Energy savings of 20% have been observed after automation. Sewage treatment plants have also been placed under computer control with ultimate process optimization and energy conservation as major goals.

Industrial Boilers

Energy management is increasingly becoming the "norm" for energy-intensive industries using large amounts of electrical, thermal and mechanical energy. Evidence indicates that a 5% improvement in industrial boiler efficiency can readily be achieved with computer control.

REFERENCES

ADL (1979) "State of the Art Review of Computer Control in the Steel Industry: A Report of the U.S. Department of Energy," IDO-1570-T21, Arthur D. Little, Inc.

Allen R. (1979) "Busy Robots Spur Productivity," *IEEE Spectrum* (September), pp. 31-36.

Anker-Johnson (1980) "IPAD: As It Relates to the Utilization of CAD/CAM in General Motors," General Motors Corporation.

Arneson, D. A. (1979) "Webb Forges Ahead in Material Handling," *Iron Age* (May).

Arneson, D. A. (1980) "Michigan Seamless Stays Put – And Expands," *Iron Age* (May 5), pp. MP26-MP31.

Arrowsmith, C. (1978) "Change to High Level Language Aids Batch-Mixing System Payback," *Control Eng.* (February), pp. 113-114.

Baensch, G., R. Lanbeck and H. Scharpenberg (1980) "Electrical Equipment for a Communal Sewage Works," *Brown-Boveri Rev.* 67:343-347.

Bailey, S. J. (1980) "Stream Analysis '80; System Concepts Gain," *Control Eng.* (June), pp. 71-76.

Bailey, S. J., and K. Pluhar (1979) "Flexible Manufacturing Systems, Digital Controls, and the Automatic Factory," *Control Eng.* (September), pp. 59-64.

Baker, A. T., and R. J. Rahoi (1981) "Programmable Controllers: Systems of Choice for Power Plant Wastewater Treatment," *InTech* (September), pp. 51-54.

Battelle (1975) "Evaluation of the Theoretical Potential for Energy Conservation in Seven Basic Industries," PB244772, Columbus, OH.

Battelle (1977) "Survey of the Applications of Solar Thermal Energy Systems to Industrial Process Heat: Vol 2: Industrial Process Heat Survey, January 1977," prepared for Energy Research and Development Administration.

Beardsley, C. W. (1971) "Computer Aids for IC Design, Artwork, and Mask Generation," *IEEE Spectrum* (September), pp. 63-79.

Beecher, R. C. (1979) "Assembly Robots," paper presented at the 43rd Annual Machine Tool Forum, Sponsored by Westinghouse Electric Corporation, Pittsburgh, PA, June 4-6.

Bennett, K. W. (1979) "Die Casting's New Era: Manufacturing Systems," *Iron Age* (May).

Branch, G. H. (1976) "Cold End Automation in the Flat Glass Industry," *Glass Technol.* 17(1):19–20.

Brodman, M. T., and C. L. Smith (1976) "Computer Control of Batch Processes," *Chem. Eng.* (September), pp. 191–198.

Buchmann (1978) "Use of Process Computers for Temperature Control of Reheating Furnaces," paper presented at the Seminar on the Economic and Technical Aspects of the Application of Computer Techniques in Iron and Steelmaking Processes, Ostrava, Czechoslovakia, June 26–July 1.

Bungay, H. R. (1980) "Inexpensive Computers Aid Continuous Fermentation," *Chem. Eng. Prog.* (April), pp. 53–54.

Business Week (1980a) "Technology, Robots Join the Labor Force" (June 9), pp. 62–76.

Business Week (1980b) "U.S. Auto Losing a Big Segment of the Market Forever" (March 24), pp. 78–88.

Bylinski, G. (1981) "A New Industrial Revolution is on the Way," *Fortune* (October), p. 106.

Cathey, P. (1979) "CF&I Pulls Out the Stops in Rail-Making at Pueblo," *Iron Age* (May), pp. MP-27,MP-28.

Challand, T. B., R. W. Colbert and C. K. Venkatesh (1981) "Computerized Heat Exchanger Networks," *Chem. Eng. Prog.* (July), pp. 65–71.

Cichelli, J. A. (1979) "The Maintenance and Problem Aspects of Microcomputers in the Steel Industry," *Inst. Assoc. Am.* (44A):1243–1249.

Clapp, N. W. (1979) "Three Laws for Roboticists: An Approach to Overcoming Worker and Management Resistance to Industrial Robots," Block Petrella Assoc.

Control Engineering (1981) "World's First Automated Sawmill Gets a Log up on Competition" (January), p. 35.

Craig, R. H. (1979) "Induction Heating of Slabs at McLouth Steel," *Iron Steel Eng.* (September), pp. 50–55.

Craig, J., and J. O'Hanlon (1978) "European Blast Furnace Trials with Fluxed Pellets," *Ironmaking Proceedings, Vol. 37* (Chicago, IL: Iron and Steel Society of AIME).

Dewar, R. (1979) "Machine Vision," paper presented at the 43rd Annual Machine Tool Forum, Sponsored by the Westinghouse Electric Corporation, Pittsburgh, PA, June 4–6.

Diagre, L. C., and G. R. Wieman (1977) "Computer Control of Ammonia Plants," *Chem. Eng. Prog.* (February), pp. 50–53.

DiBiano, R. (1981) "Importance of Versatile Control Strategy on Crude Unit," *Chem. Eng. Prog.* (February), pp. 56–64.

DOC (1976–1978) "Annual Survey of Manufacturers, Fuels and Electric Energy Consumed, Industry Groups and Industries," M74(AS)-4 and M76(AS)-4, U.S. Department of Commerce, Bureau of the Census, Washington, DC.

DOC (1980a) "1980 U.S. Industrial Outlook for 200 Industries with Projections for 1984," U.S. Department of Commerce, Industry and Trade Administration, Washington, DC.

DOC (1980b) "1977 Census of Manufacturers, Glass Products," MC77-1-32-A, U.S. Department of Commerce, Washington, DC.

DOC (1980c) "1977 Census of Manufacturers, Agricultural Chemicals," MC77-1-280, U.S. Department of Commerce, Washington, DC.

DOC (1980d) "1977 Census of Manufacturers, Industrial Organic Chemicals," MC77-1-28F, U.S. Department of Commerce, Washington, DC.

DOC (1980e) "1977 Census of Manufacturers, Miscellaneous Chemical Products," MC77-1-28H, U.S. Department of Commerce, Washington, DC.

DOC (1980f) "1977 Census of Manufacturers, Motor Vehicles and Equipment," MC77-137A, U.S. Department of Commerce, Washington, DC.

DOC (1980g) "1977 Census of Manufacturers, Drugs," MC77-1-28C, U.S. Department of Commerce, Washington, DC.

DOE (1979) "Computer Technology: Its Potential for Industrial Energy Conservation," DOE/CS/2123-TS, U.S. Department of Energy, Washington, DC.

DOL (1974a) "Technological Change and Manpower Trends in Six Industries," Bull. 1817, U.S. Department of Labor, Washington, DC.

DOL (1974b) "Computer Manpower Outlook," Bull. 1826, U.S. Department of Labor, Washington, DC.

DOL (1975) "Technological Change and Manpower Trends in Five Industries," Bull. 1856, U.S. Department of Labor, Washington, DC.

DOL (1979) "Technological Change and Its Labor Impact in Five Energy Industries," Bull. 2005, U.S. Department of Labor, Washington, DC.

DPRA (1976) "Energy Efficiency Improvement Targets: Food and Kindred Products Industry (SIC 20), Vol. 2, Appendix Report, Part 1," PB 269 831, Development Planning and Research Associates, Inc., Federal Energy Administration, Washington, DC.

Edgington, S. H. (1979a) "Controls in the Glass Industry and Future Automation: Part 2," *Measurement Control* 12:336–339.

Edgington, S. H. (1979b) "Controls in the Glass Industry and Future Automation: Part 1," *Measurement Control* 12:281–336.

Engelberger, J. F. (1980) "Robots and Automobiles: Applications, Economics, and the Future," Congress and Exposition, Society of Automotive Engineers, Inc., Detroit, MI, February.

Ernst, B. D. (1980) "Economic Justification for Industrial Robots," Ford Motor Company, Dearborn, MI.

Fadum, O. (1980) "A Computer Controlled Vapor Phase Digester with a Preimpregnation Vessel," *Tappi* 63(7):43–52.

Farquharson, D. C. (1979) "Microprocessor-based System for Blending Coal for Coke Ovens," Proceedings of the Conference on Weighing and Force Measurement, Brighton, England.

Farquharson, D. C., C. Kulper and G. F. J. Turner (1978) "Prototype Intelligent Ratio Controller for Coal Blending at Coke Ovens," paper presented at the Symposium on Metal Process Instrumentation, Randburg, South Africa, September 20–21.

FDC Reports (1981) "Computerized of Rx Production Facilities is Wave of Future" (July 13), p. 8.

Feder, A. (1977) "Test Results on Computer Graphics Productivity for Aircraft Parts Design and Automated Machining," Northrop Corporation, Hawthorne, CA.

Finger, T. H. (1978) "Direct Digital Control of a Glass Melter Refiner and Forehearths Using Small Computers," *Inst. Electrical Electronics Eng. Trans. Ind. Appl.* 14(1):62–69.

Flotho, G., H. Polkaehn, R. Munz and H. Munz (1980) "Process Control Systems for Installations in Drinking Water Supply Networks," *Brown-Boveri Rev.* 67:356–362.

Frost, H. C. (1980) "Can You Justify Automatically-Controlled Processes?" *Food Eng.* (April), pp. 90–91.

Fulton, R. E. (1980) "National Meeting to Review IPAD Status and Goals" (July/August), pp. 49–52.

Funck, A. (1978) "Computer Control Systems in the Luxembourg Steel Industry," International Iron and Steel Institute Presentation of Technical Papers.

Funk, G. L. (1980) "Automation Turns Energy Conservation Theory into Reality," *Chem. Eng. Prog.* (April), pp. 46–52.

Gamacke, L. D. (1968) "Continuous, On-Line Stream Analyzer for Dissolved CO_2 in Beer," *ASBC Proc.* pp. 120–131.

GAO (1980) "Slow Productivity Growth in the U.S. Footwear Industry—Can the Federal Government Help?" GFMSD 80-3, U.S. General Accounting Office, Washington, DC.

Gargione, F. (1980) "CAD/CAM in Packaging Aerospace Electronics," *Astronautics Aeronautics* (April), pp. 50–71.

Gibson, A. (1980) "The Potential for Energy Reduction in Breweries." *MBAA Tech. Quart.* 17(2):89–97.

Gregory, C. D., and F. G. Young (1979) "Automated Recycle Reactors for Catalyst Evaluation," *Chem. Eng. Prog.* (May), pp. 44–48.

Greene, A. M. (1981a) "Lockheed-Georgia, Advanced Technology," *Iron Age* (September), pp. 114–175.

Greene, A. M. (1981b) "Time Sharing: A Way to Get Started in Numerical Control," *Iron Age* (October), pp. 62–65.

Greene, A. M. (1981c) "Captive NC Builders See More Systems in Their Future," *Iron Age* (April), pp. 51–55.

Gremillion, J. A. (1979) "Computer Control Without the Computer," *Chem. Eng. Prog.* (May), pp. 37–43.

Groner, P. (1981) "Computer Aided Design of VLSI Saves Man Hours Reduces Errors," *Control Eng.* (April), pp. 55–57.

Hammett, J. L., and L. A. Lindsay (1976) "Advanced Computer Control of Ethylene Plants Pays Off," *Chem. Eng.* (November), pp. 115–120.

Harris, J. O., and J. Irvine (1978) "Process Considerations and Plant Design of Fermentation/Maturation Vessels in Modern Brewery Installations," *MBAA Tech. Quart.* 15(1):30–37.

Heer, E. (1981) "Robots in Modern Industry," *Astronautics Aeronautics* (September), pp. 50–59.

Herman, A. S. (1974) "Productivity in the Paints and Allied Products Industry," U.S. Department of Labor, Washington, DC.

Herold, H. (1980) "Metramatic—A Control System for Water Distribution Installation," *Brown-Boveri Rev.* 67:363–367.

Hill, R. M. (1978) "Stelco's New Blast Furnace at Nanticoke, Ontario," *Ironmaking Proceedings Vol. 37* (Chicago, IL: Iron and Steel Society of AIME).

Holmes, J. G. (1979a) "Automated Robot Machining System," paper presented at the Symposium on Industrial Robots, Dearborn, MI, March 13–15.

Holmes, J. G. (1979b) "Integrating Robots into a Machining System," Fall Industrial Engineering Conference AIIE Proceedings, Houston, TX, pp. 247–256.

Honeywell (1980) "Human Factors Affecting ICAM Implementation," Fourth Interim Technical Report.

Hope, V. E., and K. Donlan (1978) "Closed Loop Baking," *Food Eng.* (February), pp. 78–82.

Horbal, M. T., and D. Derrick (1978) "Microcomputer Systems Controls Steel Angle Fabrication," *Control Eng.* (June), pp. 69–71.

Hutchinson, G. K. (1979) "Flexible Manufacturing Systems in the United States, Automation in Manufacturing," JACC P. 743, University of Wisconsin—Milwaukee, Milwaukee, WI.

IEEE (1981) *Proceedings of the IEEE, Special Issue on Computer-Aided Design,* Institution of Electrical and Electronics Engineers.

Iizuka, M., H. Yoshida, K. Matsumura, K. Sano and N. Shinoda (1979) "NKK Thermo Camera System for Furnace Top and Its Application to Blast Furnace Operation," paper presented at the 62nd National Open Hearth and Basic Oxygen Steelmaking Conference and 38th Ironmaking Conference, March.

Iron and Steel Engineer (1979) "Minicomputer Locates Structural Flaws in Real-Time" (September), p. 59.

Iron Age (1980) "Stelco's New $1 Billion Plant Worth the Sweat," (May 5), pp. MP-17,MP24.

Jandorf, B. M. (1981) "Digital Control: What It Meant in One Boiler House," *InTech* 28(7):55-57.

Johannes, E. J. (1980) "CAD/CAM-Practical Tool for General Aviation," SAE Tecnical Paper Series, Society of Automotive Engineers, Inc., West Coast International Meeting, Los Angeles, California, August No. 800876.

Kamii, N. (1976) "Fully Computerized Bar Rolling Mill," *Iron Steel Eng.* (December).

Kennedy, J. P. (1975) "Tighter Process Design via Computer Control," *Chem. Eng.* (March), pp. 54-60.

Kern, D. W., D. M. Balla and R. J. Reinbold (1980) "Experience Using Seensor Lance on BOF's at Bethlehem Steel," *Iron Steel Eng.* (March), pp. 31-36.

King, H. S. (1979) "A Total Integrated Systems Approach to Computer Aided Design Computer Aided Manufacturing," paper presented at the CAD/CAM VII, Autofact II Conference, Society of Manufacturing Engineers, Detroit, MI.

Koekebakker, J. (1980) "The Time May Soon Arrive to Put a Robot in Your Shop," *Can. Machinery Metalworking* (January).

Kompass, E. J. (1979) "The Long-Term Trends in Control Engineering," *Control Eng.* (September), pp. 53-55.

Laird, W. W., S. J. Palko and R. J. Marcouiller (1980) "Ambridge—A Fully Computerized Seamless Tube Mill," *Iron Steel Eng.* (January), pp. 69-73.

Larsen, R. J. (1980) "Japan Exhibits Variety Among Machine Tools Firms," *Iron Age* (May), pp. 85-87.

Larsen, R. J. (1981) "Does Adaptive Control Still Promise Improved Productivity?" *Iron Age* (July), pp. 57-69.

Latour, P. R. (1976) "Energy Conservation via Process Computer Control," *Chem. Eng. Prog.* (April), pp. 76-81.

Lavigne, J. R. (1977) *An Introduction to Paper Industry Instrumentation* (Miller Freeman Publisher).

Lavigne, J. R. (1979) *Instrumentation Applications for the Pulp and Paper Industry* (Miller Freeman Publisher).

Lenard, M., and J. Bluestein (1980) "On Board Microprocessors and the Evolution of the Automobile," MP80W00013, The MITRE Corporation, McLean, VA.

Lerner, J. E. (1981) "Computer-Aided Manufacturing," *Inst. Electrical Electronics Eng. Spectrum* (November), pp. 34-39.

Lida, N., H. Miura, S. Moriya, A. Sato and J. Miyoshi (1978) "Advanced Automation on the New Plate Mill at Mizushima Works," *Iron Steel Eng.* (October).

Lockley, R. (1979) "A New Approach to Paper Mill Drives," Institution of Electrical and Electronics Engineers Annual Pulp and Paper Industry Technological Conference, March 15–18, 1979, Pittsburgh, PA, pp. 45–53.

Loibl, J. M. (1978) "Computer Control of Glass Forming and Tempering," *Inst. Electrical Electronics Eng. Trans. Ind. Appl.* IA-14(1):48–57.

Long, L. C. (1978) "Computer Based Process Control in the North American Steel Industry 1978," paper presented at the American Iron and Steel Institute General Meeting, New York, Technical Session, May 25.

Lucet, M., J. P. Toulet, J. L. Gaultier and J. F. Michel (1980) "Optimize Cold Cycles," *Hydrocarbon Processing* (July), pp. 85–93.

Luton, A. R. F. (1978) "Energy Economization Industry — The Role of Process Control," *Brown-Boveri Rev* 11:716–723.

Lutz-Nagey, R. (1976) "The Race Towards Programmable Control," *Production Eng. Mag.* (December), pp. 51–58.

Macedo, F. X., R. D. Glatt, H. Brown and A. V. Simpson (1977) "Development and Operation of a Computer-Based Reheating Furnace Control System," AIME Symposium on Automation in the Iron and Steel Industry, Atlanta, GA.

Malkiel, C. (1978) "Automation Outlook for Semiconductor Fabrication," *Circuits Manufac.* 18(9):38–41.

Malone, L. (1978) "U.S. Steel's New No. 8 Blast Furnace at Fairfield Works," *Ironmaking Proceedings, Vol. 37* (Chicago, IL: Iron and Steel Society of AIME).

Manufacturing Engineering (1980a) "Boeing Marine Adopts Computer-Controlled Welding" (May), pp. 77–78.

Manufacturing Engineering (1980b) "A Different Type of Control — Industrial Control Microcomputers" (May).

Manufacturing Engineering (1980c) "A New Generation of Computer Numerical Controls" (May), pp. 44–49.

Manufacturing Engineering (1980d) "Evolution of Microelectronics Technology" (May), p. 57.

Manufacturing Engineering (1980e) "Status Report on Programmable Controllers" (May), pp. 49–54.

Manufacturing Engineering (1980f) "System Lists Downtime Causes, Production Information" (May), pp. 71–74.

Manufacturing Engineering (1980g) "Computer and Automated Systems Association of SME, Computer Aided Design and Integrated Manufacturing" (January).

Manufacturing Engineering (1980h) "New CAD Developments" (May), pp. 81–83.

Manufacturing Engineering (1980i) "Computer-Aided Factory Management Promises Increased Productivity," p. 83.

Manufacturing Engineering (1980j) "Vertical Turning Centers Triple Productivity" (May), pp. 89–90.

Marsh, P. (1980a) "Robots See the Light," *New Scientist* (June 12), pp. 238–248.

Marsh, P. (1980b) "Towards the Unmanned Factory," *New Scientist* (July 31), pp. 373–377.

Marsh, P. (1980c) "Britain Grapples with Robots," *New Scientist* (April 24), pp. 183–187.

Martyn, G. W. (1974) "The People Computer Interface in a Capsule Molding Operation Drug," *Develop. Commun.* 1(1):39–43.

Mauderli, A., and D. W. T. Rippin (1980) "Scheduling Production in Multipurpose Batch Plants: The Batchman Program," *Chem. Eng. Prog.* (April), pp. 37–45.

McCool, J. (1979) "Microprocessors in Control of Robots," *Electronics Power* (November/December), pp. 796–799.

Metsavirta, A., and M. Leppanen (1980) "Improved Probability of Thermomechanical Pulping Using Computer Control and Effective Heat Recovery," *Tappi* 63(7):37–41.

Milk Industry (1978) "Edewecht: The World's Most Automated Cheese Plant."

Miyazaki, Y., and A. Ito (1978) "Integrated Information and Control System in the Japanese Steel Industry," International Iron and Steel Institute, Presentation of Technical Papers, Vol. 1, April.

Modern Machine Shop (1980) "NC/CAM Outline."

Monteith, W. (1978) "American Steel Industry Process Control Trends," Proceedings, AIIE 1978 Spring Annual Conference.

Moore, C. E. (1979) "Are We Really Ready for VLSI[2]?" Proceedings of the Caltech Conference in Very Large Scale Integration, California Institute of Technology.

Munson, G. E. (1978) "Robots Quietly Take Their Places Alongside Humans on the Production Line to Raise Productivity — and Do the 'Dirty Work'," *Inst. Electrical Electronics Eng. Spectrum* (October), pp. 66–70.

New Scientist (1980a) "Computers Wheel onto Factory Floor" (August 7).

New Scientist (1980b) "Slow March for Britain's New Robot" (August 7), p. 455.

New Scientist (1980c) "France Leaps Forward with Robots" (July 31), pp. 369–370.

Nitzan, D., and C. A. Rosen (1976) "Programmable Industrial Automation," *Eng. Appl. Sci.* (July).

Optical Spectra (1981) "Giving Vision to Tomorrow's Robots" (September), pp. 35–40.

ORAU (1980) "Industrial Energy Use Data Book," ORAU-160, Oak Ridge Associates Universities.

ORVSC (1970) "Orsanco Quality Monitoring," Ohio River Valley Sanitary Commission.

Ouellette, R. P., R. S. Greeley and J. W. Overbey (1975) "Computer Techniques in Environmental Science," *Petrocelli/Chater.*

Persigehl, E. S., and J. D. York (1979) "Substantial Productivity Gains in the Fluid Milk Industry," *Monthly Labor Rev.* (July), pp. 22–27.

Powell, R. P. (1979) "Computerized Process Controls on Batch Digesters Save Energy," *Tappi* 12(12):21–23.

Ray, C. T., and W. L. Wilbern (1979) "The Operation of a Two-Furnace Ferrosilicon Plant Under Process Computer Control," AIME Electric Furnace Conference, Detroit, MI.

Ritchey, K. J., F. B. Canfield and T. B. Challand (1976) "Heavy-Oil Distillation via Computer Simulation," *Chemical Eng.* (August), pp. 79–88.

Ritter, J. C., and K. J. Mundy (1971) "A Process Control System for Production Diesel Engine Testing," National Combined Fuels and Lubricants, Powerplant and Truck Meetings, Society of Automotive Engineers, St. Louis, MO, October.

Rolner, R. (1970) "Continuous Automatic Control of Carbonation in Beer Steam," *ASBC Proc.* pp. 111–117.

Schenstrom, F., and R. Williams (1976) "Computer Aided Design, MOS Course — Part 2," *Electronic Eng.* (March), pp. 70–77.

Shiraiwa, T., K. Saski, T. Suzuki, S. Kobayashi and K. Matsunaga (1978) "Monitoring Method and Repairing Method for Erosion of Blast Furnace Lining," *Ironmaking Proceedings, Vol. 37* (Chicago, IL: Iron and Steel Society of AIME).

Siebergling, D. A. (1974) "Computer Controlled Digital Separating/Standardizing/Blending of Fluid Milk Products," *Am. Dairy Rev.* (January), pp. 22,24,46–47.

Skrokov, M. R. (1976) "The Benefits of Microprocessor Control," *Chem. Eng.* (November), pp. 133–139.

Smoluk, G. (1978) "Extruder Productivity: Big Gains in 5 Years," *Plasticworld* (July), pp. 48–53.

Spanbauer, J. P. (1980) "How Advanced Boiler Control Saves Energy," *Tappi* 63(7):29–31.

Spellman, R. A., and J. B. Quinn (1975) "Computer Control of Batch Reactors," Proceedings of the 2nd Joint Spring Conference, Inst. Soc. of America Montreal, Quebec, Canada.

Stauffer, R. N. (1981) "Flexible Manufacturing System, Bendix Builds a Big One," *Manufac. Eng.* (August), pp. 92–93.

Steele, K. A. (1980) "Optimization of Energy in Water Distribution Systems by Effective Measurement and Control," *Brown-Boveri Rev.* 67:350–355.

Stewart, E. L. (1978) In: *US Department of Energy Voluntary Business Energy Conservation Program Progress Report No. 6,* DOE C500118/6, Washington, DC.

Sugarman, R. (1980) "The Blue-Collar Robot," *Inst. Electrical Electronics Eng. Spectrum* (September), pp. 53–57.

Suzuki, G., M. Mizuno, M. Higuchi and T. Matsushita (1978) "Development of an Automatic Computer Control System for Coke Over Operation," *Trans. ISIJ* Vol. 18.

Takeuchi, H., T. Torigoe, T. Suzuki and K. Kise (1977) "Total On-Line Production Control System of Plate Mill," *Trans. ISIJ* Vol. 17.

Tanaka, S., H. Maeda, K. Taguchi, I. Tsuboi and A. Ozeki (1977) "Automatic Blowing Control of BOF," *Iron Steelmaker* (August).

Tappi (1979) "MODO Gets Good Results with Computer Controlled Vapor Phase Kamyr," 62(9):17–18.

Thome, R. J., M. W. Cline and J. A. Grillo (1979) *Chem. Eng. Prog.* (May), pp. 54–60.

Thompson, L. L. (1978) "Computer Control = Predictable Uniformity = Profits," *Pima* 60:19–23.

Thurston, C. W. (1979) "Experience with a Large Distribution Control System," *Control Eng.* (June), pp. 61–65.

Time (1981) "Look, No Hands, Brave New World at the Factory" (November 16), p. 127.

USAF (1979) "ICAM Program Prospectus," U.S. Air Force.

USAF (1980a) "ICAM Robotic System for Aerospace Batch Manufacturing — Task A Technical Report, Vol. 1," AFWAL-TR-80-4042, U.S. Air Force.

USAF (1980b) "ICAM Robotics Application Guide, Technical Report, Vol. 2," AFAWL-TR-80-4042, U.S. Air Force.

Vitullo, M. (undated) "The Food Industry, Industrial Energy Use Data Book," pp. 10-1 to 10-31.

Wallace, J. J. (undated) "An Overview of U.S. Manufactured Robots and Applications," Prab Conveyors, Inc., Robot Institute of America, pp. 320–326.

Weems, S. (1979) "Computer Control Stabilizes Ammonia Operations," *Chem. Eng. Prog.* (May), pp. 64–67.

Wheeler, V. (1977) "MSD Unveils First Computer-Controlled Drug Plant," *D&CI* (December), pp. 48–50.

Wilder, P. S. (1978) "The Productivity Trend in the Soaps and Detergent Industry," U.S. Department of Labor, Washington, DC.

Williams, T. J. (1980a) "Hierarchical and Distributed Control Systems for Steel Mill Applications," *Iron Steel Eng.* (April), pp. 33–38.

Williams, W. (1980b) "Fully Automated Coating Kitchens for the World's Paper Mills," *Paper* 193(5):35–41.

Williams, J. D., and R. D. Rogers (1981) "Getting Our Feet Wet in Digital Boiler Control: A Utility Case History," *InTech* (July), 28(7):49–52.

Wilson, R. J. (1978) "No. 7 Blast Furnace — Inland Steel Company," *Ironmaking Proceedings, Vol. 37* (Chicago, IL: Iron and Steel Society of AIME).

Windle, M. R., and T. E. Chappell (1981) "A Computer-Based Centralized Control and Supervisory System for an Effluent Treatment Plant," *Chem. Eng.* (371):377–383.

Wisnosky, D. E. (1979) "Computer Integrated Manufacturing," Proceedings of a Joint DOD-Industry Manufacturing Technology Workshop, September.

Wright, N. O., and J. Powers (1980) "Catalyst Loss Reduced by 21% with Process Control System," *Chem. Processing* (November), pp. 80–81.

Yost, C. C., C. R. Curtis and C. J. Ryskamp (1980) "Advanced Control at Wycon's Ammonia Plant," *Chem. Eng. Prog.* (April), pp. 31–36.

Zadarnowski, J. H. (1980) "CADD on the F-18 Program," *Astronautics Aeronautics* (March), pp. 48–57.

Zagorzycki, P. E. (1979) "Automatic Control of Conveyor Dryers," *Chem. Eng. Prog.* (April), pp. 50–56.

Zeiser, R. N. (1973) "Modernization and Manpower in Textile Mills," Reprint 2893, U.S. Department of Labor, Washington, DC.

BIBLIOGRAPHY

ADL (1975) "Steel and the Environment—A Cost Impact Analysis," Arthur D. Little, Inc., report to the American Iron and Steel Institute.

ADL (1978) "Research, Development and Demonstration for Energy Conservation: Preliminary Identification of Opportunities in Iron and Steelmaking," Arthur D. Little, Inc., NTIS SAN/1692-1, U.S. Department of Energy.

AISI (1978) "Annual Statistical Report, 1977," American Iron and Steel Institute.

American Machinist (1978) "Computers in Manufacturing" (April).

Andreiev, N. (1981) "Data Highways Led Microcomputer Peripherals to Industrial Control," *Control Eng.* (January), pp. 51-55.

Attiyate, Y. (1979) "Computer Use in Sugar Production," *Food Eng. Int.* (April), pp. 26-27.

ASME (1980) "System Lists Downtime Causes, Production Information," Society of Manufacturing Engineers, pp. 71-72.

Backman, J., J. Skogberg and H. Gedin (1980) "SKF Steel's Bar Bill at Hallefors—A New Tool for Bar Production," *Iron Steel Eng.* (May), pp. 45-52.

Barnden, M. J., and D. Bower (1978) "Computers in the Paper Industry," paper presented at the Tappi Papermakers Conference, Atlanta, GA.

Barnebey, C. (1978) "Quick Installation is a Key to Weyerhaeuser's Control System," *Pima* (December), p. 13.

Bejczy, A. K. (1980) "Sensors, Controls, and Man-Machine Interface for Advanced Teleoperation," *Science* 208(4450):1327-1335.

Benedetti, G., and W. E. Brown (1973) "Sensor Based Computer System for Management and Control," *ISA FID* 736416, pp. 119-123.

Bennett, B. W. (1980) "Automation Trickles Down to Small Farm Equipment," *Iron Age* (June), pp. 46-51.

Bertha, R. W. (1974) "How to Succeed in Automatic Assembly: Part 2—Capital Investment Justification," *Assembly Eng.* (March), pp. 12-16.

Birk, J., R. Kelly and L. Wilson "Acquiring Workpieces: Three Approaches Using Vision," University of Rhode Island, Department of Electrical Engineering, Kingston, RI, pp. 724-733.

Bowers, D. M. (1976) "Mini/Micro Computers in the Automotive Industry, The Microprocessor Controlled Automobile," *Mini-Micro Syst.* (September), pp. 44-51.

175

Brandley, J. (1980) "Tooling and Production," CDM Division, National Computer Systems, Minneapolis, MN.
Brody, H. D., and R. A. Stoehr (1980) "Computer Simulation of Heat Flow in Casting," *J. Metals* pp. 20–27.
Bullivant, K. W. (1973) "Digital Weigh Feeding for Continuous Processing," Instrument Society of America, pp. 111–118.
Business Week (1978) "Why the U.S. is Lagging in Automation" (June 5), pp. 62B–62P.
Business Week (1979) "UAW Fears Automation Again" (March 26), pp. 94–95.
Bylinsky, G. (1979) "Those Smart Young Robots on the Production Line," *Fortune* (December 17), pp. 90–96.
Cameron, A. M. (1978) "Production of Coke for a Large Blast Furnace," Algoma Steel, Iron and Steel Society and Society of Mining Engineers, October 2–3.
CAMI "A Long Range Plan for Community Development of Information Services to Improve Productivity in Design and Manufacturing in the World Durable Goods Industry," Advanced Technical Planning Committee, Computer Aided Manufacturing International.
CAMI (1980) "CAM-I, Framework Project 1980," PR-79-ASPP-01.3, Computer Aided Manufacturing International, Inc., Arlington, TX.
Casa, P. D., R. Barbieri and E. Aibertini (1980) "Better Batch Process Control," *Hydrocarbon Processing* (March), pp. 101–105.
Chemical Engineering (1981) "Combustion Controller Optimized Boiler Performance" (February), pp. 49–51.
Chemical Week (1980a) "Research That Assembly 'Worker' May Be a Robot" (March 26).
Chemical Week (1980b) "The Search for Balance Between Man and Machine" (July 16), pp. 55–62.
Cloud, F. E., and D. L. Roig (1977) "Length Measurement and Cut-out Computation of Hot Bullets," *Iron Steelmaker* (April).
Cocheo, S. (1981) "How to Evaluate Distributed Computer Control Systems," *Hydrocarbon Processing* (June), pp. 95–106.
Collin, P. H., and H. Stickler (1980) "ELRED—A New Process for the Less Expensive Production of Liquid Iron," *Iron Steel Eng.* (March), pp. 43–45.
Cone, C. (1980) "Microcomputer-Assisted Control System for Continuous Rolling Mills," *Iron Steel Eng.* (May), pp. 63–64.
Control Engineering (1979) "The Configurations of Process Control: 1979" (March), pp. 43–54.
Coombs, R., and K. Green (1980) "Slow March of the Microchip," *New Scientist* (August 7), pp. 448–450.
Critchlow, R. V. (1977) "Technology and Labor in Automobile Production," *Monthly Labor Rev.* (October), pp. 32–35.
Cullen, C. W., and G. P. Petrus (1980) "Generation III Hot Strip Mill Automation System," *Iron Steel Eng.* (November), pp. 25–30.
Cunningham, C. R. (1974) "Process Control Techniques in Airplane Manufacturing," National Aerospace Engineering and Manufacturing Meeting, Society of Automotive Engineers, San Diego, CA.
Decker, A. (1978) "Process Computer Control," International Iron and Steel Institute Presentation of Technical Papers.
DOC (1977) "Census of Manufacturers, Office Computing and Accounting Machines," MC77-1-35F, Washington, DC.

Doering, G. I. (1981) "Computer Control for Discrete Production Processes," *InTech* 28(9):81–82.

DOL (1976) "Technological Change and Manpower Trends in the Industries," Bull. 1856, U.S. Department of Labor, Washington, DC.

DOL (1977) "Technological Change and Its Labor Impact in Five Industries," Bull. 1961, U.S. Department of Labor, Bureau of Labor Statistics, Washington, DC.

Economist (1979) "Business Science and Technology, Britain Needs to Use More Robots" (November 10).

Elias, S. (1979a) "Material Handling: Computers Take Control," *Food Eng.* (September), pp. 95–96.

Elias, S. (1979b) "How PC's Can Help You," *Food Eng.* (October), pp. 89–94.

Engelberger, J. F. (1979) "Robotics 1983," *Robotics Today* (Fall).

Estes, V. E. (1978) "An Organized Approach to Implementing Robots," General Electric Company, Manufacturing Engineering Consulting and Applications Center.

Evanson, C. E., and R. W. Bolz (1977) "Automating the Plant for Increased Productivity," *Consulting Eng.* 48(9):76–82.

Farmer, A. R. (1978) "Energy Economization in Industry – The Role of Process Control," *Brown-Boveri Rev.* 11:716–723.

Farrall, A. W., L. C. Elkins and D. A. Seiberling (1979) "Automation and Computer Systems," *Food Eng. Syst.* 2:321–338.

Felix, M. P., and W. E. Simpkin (1979) "Highlight 1979, Design Engineering," *Astronautics Aeronautics* (December), pp. 47–50.

Flynn, J. E. (1978) "Uddeholm Mill Uses Computer as a Market Wedge," *Paper Trade J.* (April 1), pp. 25–29.

Food Engineering (1979a) "'Family-Style' Firm Automates Operation and Retains Batch-Type Flexibility" (May), pp. 168–171.

Food Engineering (1979b) "Brewhouse Computer Control Increases Output" (February), pp. 66–67.

Food Engineering (1980) "Random-Weight Cheese Packager Goes UPC: Retailers and Consumers Benefit" (July), pp. 134–135.

Froberg, F. (1971) "Adaptive Control and Automobile Manufacture," Automotive Engineering Congress, Society of Automotive Engineers, Detroit, MI, January.

Garde, J. L. (1978) "Now On-Stream Automatic Broken-Case Picking," *Modern Mat. Handling* (July), pp. 65–69.

Garen, E. R. (1980) "Microcomputers: The Ubiquitous Control Element," *Control Eng.* (June), pp. 53–54.

Geier, J. (1979) "What a Manufacturing Equipment Producer Sees Ahead for Improving Productivity," paper presented at the Conference on Manufacturing Productivity Solutions, Sponsored by the Society of Manufacturing Engineers and the U.S. Chamber of Commerce, Washington, DC, October.

Gettleman, K., and M. E. Merchant (1978) "Computers Advancing World, Manufacturing Technology," *Modern Machine Shop* (51):123–129.

Gibson, A. (1980) "The Potential for Energy Reduction in Breweries," *MBAA Tech. Quart.* 17(2):89–97.

Goble, W. (1981) "Understand Microcomputers," *Hydrocarbon Processing* (June), pp. 111–114.

Gorin, J., and L. Stern (1980) "The Case for Board-Level Microcomputers," *Mini-Micro Syst.* (November), pp. 81–88.

Govsievich, R. E., et al. (1978) "Determining the Economic Effectiveness of Industrial Robots," *Machines Tooling* 49(8):10–12.

Greene, A. M. (1978) "NC-CAD/CAM REPORT, A Management Guide to Computer-Integrated Manufacturing," *Iron Age* (December 18).

Greene, A. M. (1980) "DNC: The Concept Becomes a Reality," *Iron Age* (November), pp. 51–64.

Grignet, J. (1981) "Microprocessor Improves Wool Fiber-Length Measurements and Extends the Application," *Textile Res. J.* 5113:174–181.

Guerens, P. (1977) "Automatic Furnace Control," *Glass Technol.* 18(2):79–81.

Hahn, C. W. (1977a) "Process Control and Instrumentation in Breweries — Part 1," *MBAA Tech. Quart.* 14(1):59–69.

Hahn, C. W. (1977b) "Process Control and Instrumentation in Breweries — Part 2," *MBAA Tech. Quart.* 14(2):87–93.

Hammett, J. L. (1980) "CRT Based System: What's in It for the Operator?" *Hydrocarbon Processing* (July), pp. 135–141.

Harvey, D. (1979) "When the Robots Take Over — What's Left?" *Chief Executive* (December), p. 43–47.

Harvey, A., A. F. Ogg and J. M. Hugill (1978) "Factors Affecting the Choice of an On-strand Cooling Sinter Plant," *Ironmaking Proceedings Vol. 37* (Chicago, IL: Iron and Steel Society of AIME).

Hawkinson, J. M. (1978) "The Microprocessor Based Controller in Industry, Is It Reliable?" *Inst. Electrical Electronics Eng. J.* pp. 64–68.

Hinrichsen, E. N. (1976) "Hot Strip Mill Runout Table Cooling — A System View of Control, Operation, and Equipment," *Iron Steel Eng.* (October).

Hintz, O. E., J. H. Sullivan and R. C. Van Sarys (1978) "Machine Tool Industry Study, Final Report," U.S. Army Industrial Base Engineering Activity, Rock Island, IL.

Hoare, P. A. (1978) "The Benefits of Automatic Process Control," 31(4):217–219.

Hohn, R. E. (undated) "Robot Path Control by Off-Line Computer," Condec Corporation U.S.A., pp. 338–345.

Hulburt, D. A., and R. Albaugh (1979) "Risk Analysis of an Automated Assembly Line Operation," Congress and Exposition, Society of Automotive Engineers, Inc., Detroit, MI, March.

Hutchinson, G. K. (1977) "Flexible Manufacturing Systems in Japan," National Technical Information Service, Springfield, VA.

IDF (1977) "Annual Bulletin, Safety and Reliability of Automated Dairy Plant," International Dairy Federation, pp. 1–25.

Inagaki, S. "Problems Awaiting Solutions of Servomechanisms of Industrial Robots," Nagoya Municipal Industrial Research Institute, Japan, pp. 558–565.

InTech (1979) 26(5):32.

IPBI (1975) "Corrugator Automation," International Paper Board Industry, pp. 50–51.

IRD "Robots to Multiply," International Resource Development, Inc.

Iron and Steel Engineer (1978) "Energy Savings Through Centralized Control" (September).

ISIJ (1976) *Trans. ISIJ* 16(589).

Jandorf, B. M. (1981) "Digital Control: What It Meant in One Boiler House," *InTech* 28(7):55-57.

Jenkins, D. W. (1978) "Distribute or Not? The Choice Between Distributed and Central Computing Control," *Control Eng.* (June), pp. 61-64.

Journal of Metals (1979) "Industry and Business News" (February).

Kelly, R. L. (1980) "Energy Conservation for the 80's," paper presented at State Energy Audit Impact '80, St. Louis, MO, March 31-April 2.

Kemlo, G. (1978) "Reduction of Ingot Butt Losses," *Steelmaking Proceedings, Vol. 61* (Chicago, IL: Iron and Steel Society of the AIME).

Kline, M. J. (1978) "Conserving Energy in Frozen Food Warehouse Design," *Food Eng.* (April), pp. 76-78.

Knill, B., and G. Schwind (1976) "Clark Advances Technology and Management Concepts at New Lift Truck Plant," *Mat. Handling Eng.* (November), pp. 56-63.

Kompass, E. J. (1978) "Fitting Computers to the Control Task," *Control Eng.* (June), pp. 50-51.

Krasny, C., and A. J. Harris (1976) "In-line Automatic Shearing, Stacking and Bundling Round and Shaped Products," *Iron Steel Eng.* (November).

Kuljian, M. J. (1979) "Automatic Microprocessor Control of Continuous In-Line Aluminum Deposition on Silicon Wafers," *JACC* p. 266-275.

Larsen, R. J. (1978) "Better Loading/Unloading: The Key to Productivity," *Iron Age* (December), pp. 53-56.

Larsen, R. J. (1981) "Researchers Spread News of Flexible Manufacturing," *Iron Age* (September), pp. 99-107.

Larsen, R. J., R. J. Hocken, J. Tlusty, R. V. Miskell and A. R. Thomson (1980) "Where Machines Leave Off, People Take Over. Technology," *Iron Age* (May 19), pp. 68-81.

LeCerf, B. H., G. A. Weimer and C. T. Post (1978) "Metalworking's Future Manufacturing Systems," *Iron Age* (August 28), pp. 68-118.

Lamaire, W. H. (undated) "Computer Spots Profit-Gobblers," *Food Eng.* p. 108.

Lemaire, W. H. (1978a) "Remarkable Advances in Process Control," *Food Eng.* (July), pp. 55-56.

Lemaire, W. H. (1978b) "Dairy Co-ops Unwrap New Custom Dehydrating Plant," *Food Eng.* (January), pp. 81-82.

Lloyd, M. (1975) "Computer Control of Brewhouse Grain Batching and Brew Scheduling," *MBAA Tech. Quart.* 12(4):214-221.

Long, L. C. (1977) "Steel Mill Automation, Progress and Prognosis," *Iron Steelmaker* (March).

Long, L. C., and J. H. Schunk (1979) "Applications of Hierarchical Control in the Steel Industry," Armco Inc., Middletown, Ohio, U.S.A., *JACC* pp. 638-652.

Mann, P. J. (1978) "Electronic Water Treaters: How Good Are They?" *Food Eng.* (April), pp. 95-96.

Martin, J. R. (1969) "The Analog or Digital Choice," National Farm, Construction and Industrial Machinery Meeting, Society of Automotive Engineers, Milwaukee, WI, September.

McGannon, H. E. (1971) "The Making, Shaping and Treating of Steel," United Steel Company, Pittsburgh, PA.

McGowan, M. (1978) "I/O Interfaces: A Microcomputer's Link to the Real World," *Control Eng.* (June), pp. 56–58.

McGraw-Hill (1978) *World Steel Industry Data Handbook for the United States, Vol. 1* (New York: McGraw-Hill Publishing Company), pp. 149–154.

Mechanical Engineering (1981) "CAD/CAM Loop for Automatic Machining of Large Parts" (May), p. 45.

Melville, H. J. (1981) "Programmable Controllers Improve Batch Digester Performance," *InTech* (September), pp. 57–60.

Metalworking (1980) "Concepts of Unattended Machining" (September), pp. 44–52.

Michalopoulos, D. A. "Robot Drills, Novts Parts for All Management Computers," *New Appl.* p. 130.

Michelini, R. C., P. L. Polledro and C. M. Taddei (undated) "Position Steering of Industrial Robots by Statistical Controllers," Institute of Applied Mechanics, University of Applied Mechanics, University of Genova, Department of Mechanical Engineering Lehigh University, Bethlehem, PA, pp. 618–625.

Modern Materials Handling (1980) "Industrial Robots: Better Than a Man — Sometimes!" (April), pp. 90–97.

Mollis-Mellberg (1979) "Process Computer/Automation Systems," Outokumpu, Technical Export Division.

Mori, K., and K. Sugiyama "Material Handling Device for Irregularly Shaped Heavy Works," Mechanical Engineering Research Laboratory, Hitachi, Ltd., pp. 504–511.

Morris, C. E. (1980) "The Cheese Boom, Part III," *Food Eng.* (January), pp. 125–128.

Mueller, G. E., R. C. Werner and R. L. Whiteley (1974) "Computer-Directed Plate Mill Automatic Gauge Control Systems," *Automation in the Iron and Steel Industry,* Moyan, R. P. and J. Szekeley, Eds.

Myers, W. (1976) "Key Developments in Computer Technology: A Survey," *Computer* 9(11):48–77.

Nevins, J. L., and D. E. Whitney (1978) "Computer-Controlled Assembly," *Scientific Am.* 238(2):62–74.

Nevins, J. L., D. E. Whitney and the Draper Laboratory Staff (1980) "Assembly with Robots(?) — Design Tools or Technology Question," paper presented at the Congress and Exposition, Society of Automotive Engineers, Inc., Detroit, MI, February.

New Scientist (1980a) "Computers Wheel onto Factory Floor" (August 7).

New Scientist (1980b) "Slow March for Britain's New Robot" (August 7), p. 455.

New Scientist (1980c) "France Leaps Forward with Robots" (July 31), pp. 369–370.

Newsweek (1980) "An Economic Dream in Peril" (September), pp. 50–65.

NSF (1978) "Research Needs of the Automation Field," American Automatic Control Council, National Science Foundation, Wickliff, OH, pp. 286–325.

Obrzut, J. J. (1981) "Computer Fine Tunes Structural Steel Fabrication Line," *Iron Age* (April), pp. 51–55.

O'Connell and T. N. Thorla (1980) "Modernizing a Hot Strip Finishing Mill Main Drive Control at Inland Steel," *Iron Steel Eng.* (May), pp. 45–52.

Oil and Gas Journal (1977) "Ammonia Process Control Improved" (May 30), pp. 113–114.

Ouellette, R. P., N. W. Lord and O. G. Farah (1981) "Electricity Use in the

Industry (Past, Present, Future)," MP-81-W26, paper presented at the Future of Electric Power Workshop, McLean, VA, April.

Package Engineering (1979) "Smart Checkweighters Do More Than Give Your Product a Weigh" (February), pp. 29–33.

Patel, N. R. (1970) "Computer Control of Float Glass at Ford," *Inst. Electrical Electronics Eng. Trans. Ind. Gen. Appl.* 1GA-6(4):375–377.

Paul, R. (1979) "Robots, Models, and Automation," Purdue University, West Lafayette, IN.

Perryman, R. R., and N. E. Prochaska (1972) "User's View of Process Control Computer Systems Management," National Automobile Engineering Meeting, Society of Automotive Engineers, Detroit, MI, May.

Perzley, W. "Robot Path Control by Off-Line Computer," Condec Corporation, U.S.A., pp. 338–345.

Plonka, F. E., W. M. Hancock and P. T. Sathe (1979) "Design of a Process Control System for Automotive Assembly Process," Congress and Exposition, Society of Automotive Engineers, Inc., Detroit, MI, March.

Pluhar, K. (1979) "Distributed Control in Discrete Past Manufacturing – An Overview," *Control Eng.* (September), pp. 57–58.

Post, C. T. (1980) "How Can Industry Combat Welding Fume Problem?" *Iron Age* (May 12), pp. 45–48.

Potter, R. (1977) "Applications of Industrial Robots That Can See," International Automotive Engineering Congress and Exposition, Society of Automotive Engineers, Detroit, MI, March.

Process Engineering (1981) "Instrumentation in the 80's" (April), pp. 62–65.

Rosen, C. A., and D. Nitzan (1977) "Use of Sensors in Programmable Automation," *Computer* (December), pp. 12–23.

Roundy, R. (1979) "Exotic Technologies," *Food Eng.* (February), pp. 68–70.

Rowen, H. (1980) "U.S. Auto Industry, Workers Wonder How to Cope with Japanese Problem," *Washington Post* (May 11).

Ruocco, J. J., R. W. Coe and C. W. Hahn (1980) "Computer Assisted Exotherm Measurement in Full-Scale Brewery Fermentations," *MBAA Tech. Quart.* 17(2):69–76.

Russo, J. R., and R. Bannar (1978) "Instrumentation: What the Food Industry Wants," *Food Eng.* (February), pp. 72–76.

Sandefur, M. J. (1980) "How to Implement Digital Control," *Hydrocarbon Processing* (July), pp. 115–125.

Sapakie, S. F., D. R. Mihalik and C. H. Hallstron (1979) "Drying Techniques: Drying in the Food Industry," *Chem. Eng. Prog.* (April), pp. 44–49.

Schlatter, H. G. (1980) "Microcomputer-Assisted Control System for Continuous Rolling Mills," *Iron Steel Eng.* (May), pp. 63–64.

Schumacher, J. J. (1976) "Energy Conservation Through the Utilization of Automation," Earthmoving Industry Conference, Central Illinois Section, Society of Automotive Engineers, Peoria, IL, April.

Seeman, R. C., and A. E. Nisenfield (1975) "Distillation Column Control Offers Flexibility," *Oil Gas J.* (October), pp. 57–60.

Seman, N. G. (1979) "Robots are Molding a New Image – Part 1," *Foundry Manag. Technol.* 107(5):30–40.

Shah, G. C. (1980) "Understanding Minicomputer Control Systems," *Hydrocarbon Processing* (April), pp. 15–107.

Shinskey, F. G. (1976) "Energy-Conserving Control Systems for Distillation Units," *Chem. Eng. Prog.* (May), pp. 73–78.

Shumaker, G. C. (1980) "Robotics," *In Depth* pp. 13–20.

Simms, G. J., N. D. Burns and K. Oldham (1981) "15-Interactive Computer-Assisted Design of Circular Weft-Knitting Machines," *J. Textile Inst.* (4):162–167.

Stanton, B. D. (1981) "Designing Distributed Computer Control Systems," *Hydrocarbon Processing* (June), pp. 107–110.

Steelman, D. M. (1974) "Application of a Small Computer to the Annealing Process," *Wire J.*

Stults, B. (1978) "Food Process Instrumentation and Control," *Food Technol.* (March), pp. 22,24.

Sugarman, R. (1978) "Productivity II—Electrotechnology to the Rescue," *Inst. Electrical Electronics Eng. Spectrum* Vol. 15, October.

Teresko, J. (1980) "Can Robots Integrate Manufacturing Plants?" *Ind. Week* (January 7).

Thomasson, F. Y. (1979) "Computer Control of a Large Industry Energy Computer," *Inst. Electrical Electronics Eng. Spectrum.*

Thompson, P. D. (1979) "Milking Equipment—Where Are We Headed?" *J. Dairy Sci.* 62(1):162–167.

Thompson, L. (1980) "Programmable Controllers on the Production Floor," *Food Eng.* (February), pp. 58–59.

Thompson, P. W. (1981) "A User-Oriented Microprocessor Control System," *Manufac. Chemist Aerosol News* (July), p. 47.

Turpen, B. L. (1981) "Why Don't Your Instrument Systems Operate as You'd Expected?" *InTech* (September), p. 83.

Tyreus, B. D., and W. L. Luyben (1976) "Controlling Heat Integrated Distillation Columns," *Chem. Eng. Prog.* (September), pp. 59–66.

van den Berge, H. "Modern Automation in the Food and Beverage Industry," *Sci. Ind.* (13):36–40.

Vandiver, R. L. (1981) "What Is Distributed Control?" *Hydrocarbon Processing* (June), pp. 91–94.

Villee, G. N. (1978) "The Modern Way—Automatic Inspection and Conditioning of Billets and Blooms," *Iron Steel Eng.* (May).

Wagner, F. A. (1979) "Distributed Microcomputer Control System for a Blast Furnace," Proceedings of the 1979 Joint Automatic Control Conference, American Institute of Chemical Engineers, pp. 271–275.

Weimer, G. A. (1979) "New Timken Tube Mill: A Machine in Itself," *Iron Age* (May), p. MP-25.

Wellrorn, E. W., and F. E. Lambert (1974) "A Direct Digital Control System in Instruments in the Pulp and Paper Industry," Instrumentation Society of America, Vol. 15.

Wiechec, W. J., and L. R. Albaugh (1979) "Process Analysis," paper presented at Congress and Exposition, Society of Automotive Engineers, Inc., Detroit, MI, March.

Zecca, A. R., and J. H. Schunk (1977) "A Dynamic Control Model of Box Annealing," *Iron Steel Eng.* (June).

INDEX

183